TRANSACTIONS IN INTERNATIONAL LAND MANAGEMENT

T0252849

To

Jennifer and Margit

Transactions in International Land Management
Volume 1

Edited by

ROBERT W. DIXON-GOUGH
University of East London, UK

REINFRIED MANSBERGER
Universität für Bodenkultur, Austria

LONDON AND NEW YORK

First published 2000 by Ashgate Publishing

Reissued 2018 by Routledge
2 Park Square, Milton Park, Abingdon, Oxon OX14 4RN
711 Third Avenue, New York, NY 10017, USA

Routledge is an imprint of the Taylor & Francis Group, an informa business

Publisher's Note
The publisher has gone to great lengths to ensure the quality of this reprint but points out that some imperfections in the original copies may be apparent.

Disclaimer
The publisher has made every effort to trace copyright holders and welcomes correspondence from those they have been unable to contact.

A Library of Congress record exists under LC control number: 00133623

ISBN 13: 978-0-8153-8255-3 (hbk)
ISBN 13: 978-0-8153-8258-4 (pbk)
ISBN 13: 978-1-351-20811-6 (ebk)

Contents

Introduction

Transactions in International Land Management in common with the (accompanying) book series has the aim of encouraging the study of the complex international and multidisciplinary issues involved in international land management, such as

- surveying and valuation of the land;
- concepts of environmental issues of sustainable land development;
- acquiring semantic data of the land;
- methodologies and a comparative analysis of current practices;
- management of the land and tools for planning and land management;
- land conservation;
- evolution of the landscape;
- regulation and legislation relating to land management;
- cultural, human and social issues of land management.

Both the book series and Transactions aim to make available to students, lecturers and practical experts, material that will focus on the various aspects of land management at a variety of levels. These will range from undergraduate teaching, the dissemination of practical applications, through to the dissemination of research.

The fundamental aim of the editors and the editorial board is to encourage and promote the study of land management in the broadest possible sense. Contributions are invited from practical experts, consultants and those involved in the teaching of the discipline. A second aim is to ensure that research findings are made available to a wider audience of interested readers and that research is integrated into current practice through its wider dissemination. The third aim is to make available studies based upon the themes of conferences.

Transactions in International Land Management owes its evolution to a very successful conference organised by Dr. Richard Bullard in January 1997. The conference was on International Land Management and was organised in conjunction with the Royal Institution of Chartered Surveyors. It attracted experts in land management from across the world. Although the initial intention was to publish the papers as the proceedings of the conference, it was subsequently considered that the edited and refereed papers could be reproduced as the first four volumes of a new series of *Transaction in International Land Management*.

Transactions will be published on an irregular basis although it is anticipated that a minimum of two volumes per year will be produced. It is anticipated that it will become an important 'journal' in the subject field of International Land Management. There is currently no competing journal that covers the diverse aspects of this subject area.

The following volumes of Transactions are already planned. They consist of:

TILM Volume 2.

- Contaminated land investigation and risk assessment. A case study in Portsmouth. *Clay, J. and Viney, T.*
- Analysis of the economic constraints to multiple use forest management in the Amazon region. *de Gama, Z.A.Q.P., Hoeflich, V.A. and Ggraça, L.R.*
- Riparian zones: issues of definition and management. *Jarvis, A.D. and Cook, H.F.*
- Towards the definition of a pan European integrated coastal zone management framework. *Dixon-Gough, R.W.*
- Land management in rural areas in Poland. Background to the economic transformation in the nineties. *Muczynski, A., Surowiec, S. and Zebrowski, W.*
- Was the change in land ownership necessary? A Bulgarian conundrum. *Marion, R.*
- Radical changes in rural spaces in the eastern part of Germany and Eastern Europe. *Tnöne, K.F.*
- The evaluation of land use changes in the East Thames corridor. *Home, R.*

TILM Volume 3.

- The impact of ecological demands on the process of reallotment. *Buis, A.M.*
- Land management in small island states. *Greenwood, D.W.*
- The channel tunnel link – land use problems in Kent. *Haywood, R.*
- Environmental management: the challenge for chartered surveyors. *Markwell, S. and Ravenscroft, N.*
- Property rights revisited: institutional changes and land reform. A case study in community forestry in Nepal. *Quinn, F.*
- Layout of farmland plots. *Per Sky*
- Land management for the development of rural areas – the Bavarian approach. *Stumpf, M.*
- Urban land readjustment in the development process – the influence of property owners. *Viitanen, K.*
- Management of agricultural real properties situated in town limits. *Źróbek, R. and Źróbek,S.*
- The application of terrestrial photogrammetric techniques for the monitoring of cliff faces. *Hobbs, K.F.*

TILM Volume 4.

- The cadastral system of Kiev city: managing real estate registration and asset valuation. *Deakin, M.*
- Implementing negotiated land reform: the case of Columbia. *Heath, J.*
- Going beyond Shibboleths: the place of land tenure in urban management in Kenya. *Kiamba, C.M.*
- Serving the landless: the Grameen way. *Momen, M.A.*
- Citizenship, society and the management of land. *Ravenscroft, N.*
- Would a registry map hang comfortably in a round mud hut? A register of titles for Zimbabwe's communal areas: philosophy and technical considerations. *Törhönen, M.-P. and Greenwood, D.P.*

- Development of land for new urban areas: in search of a new approach. *de Wolff, H.*
- Managing the land under cloves and coconuts: the Zanzibar experience. *Sulaiman, M.S.*
- Impact of spatial distribution of land contamination on property investment appraisal. *Kennedy, P.J., Nathanail, C.P., Abbachi, A. and Martin, I.D.*
- The importance of Global Positioning Systems in the mapping and management of rural properties. *Dare, P.*

The first book of the series, *Land Reform and Sustainable Development,* was published at the beginning of November, 1999 and the texts currently being prepared for publication include:

Land Consolidation and Rural Development
European Coastal Zone Management
Communal Land Ownership: Remnant of the Past?

If you are interested in submitting a paper for publication in the *Transactions of International Land Management*, please contact either the editors or any member of the Editorial Board.

The provisional Editorial Board for this, the first volume of Transactions, consists of:

This Board will be expanded as the series develops.

With each volume of Transactions, a brief 'portrait' will be given of some of the members of the Editorial Board.

Professor Allan Brimicombe graduated in geography from the University of Sheffield in 1976 and subsequently spent 12 years as an engineering geomorphologist with Binnie and Partners working on projects in Hong Kong, China and Malaysia. He joined Hong Kong Polytechnics in 1988 and became the founding Head of the Department of Land Surveying and Geo-Informatics. He returned to the UK in 1995 to become the first Head of the newly combined, School of Surveying at the University of East London, being awarded the Chair in 1999.

He has published extensively and his principal research interests lies in the operational aspects of GIS.

Frank Byamugisha graduated from Makerere University in 1977 with a Bachelors Degree in Agriculture (Agricultural Economics). This was followed by Masters Degrees in Agricultural Development Economics, from the Australian National University (1980), and Surveying from the University of East London (1999). Between 1980 and 1983, he worked as a Senior Economist and later Assistant Secretary for Economic Policy in the Department of Finance of the Government of Papua, New Guinea. Since September 1983, he has worked at the World Bank, where he is currently a Principal Operations Officer and Chairman of the World Bank's Land Policy and Administration Thematic Group.

His day-to-day work has largely been confined to work on land projects across East Asia and the Pacific Region. To ensure that he remains updated in other regions of the world, he has started to get involved in land-related projects in Bulgaria and market-assisted land reforms in Latin America.

Dr Hadrian Cook is a Senior Lecturer in Hydrology. A member of the Chartered Institute of Water and Environmental Management and the British Society of Soil Science. He teaches most aspects of Hydrology and water policy together with aspects of environmental history. His main research interests are: groundwater protection policy development, soil water conservation and the management and operation of historic water management systems such as watermeadows and grazing marshes. In addition to his research publications, he has written and edited books concerned with water policy and management.

Dr Mark Deakin is a Senior Lecturer and Teaching Fellow in the School of the Built Environment, Napier University. He has also been responsible for developing under-graduate and post-graduate programmes in land and property management. The undergraduate programmes include BSc (Hons) degrees in Estate Management and Planning and Development Surveying. At post-graduate level the programmes include an MSc and MBA in Property and Construction Management.

He has carried out research into the development of land management for the past five years. The research projects in question include those on contemporary land and property management funded by the Royal Institution of Chartered Surveyors, British Know-How Fund and Overseas Development Agency. As a partner in BEQUEST (Building Environmental Quality Evaluation for Sustainability through Time), he is responsible for reviewing the assessment methods currently available for the evaluation of sustainability. This research project is funded by the European Commission. He is a regular contributor to property journals specialising in the field of management science and has recently edited a book entitled: *Local Authority Property Management: Initiatives, Strategies, Re-organisation and Reform.*

Robert Dixon-Gough has been a member of staff of the School of Surveying since 1974 and is currently a Senior Lecturer. Previously, he was a cartographer who specialised in water resource planning. He became actively interested in remote sensing in 1980 and co-authored *Britain from Space* in 1985. His main research interests are: practical applications of remote sensing, particularly in the field of coastal zone management and assessing land cover changes; anthopogenic effects upon coastal regions; land information systems for the purpose of land registration and land reform; land and environmental conservation, and rural land use and planning.

Mr Dixon-Gough is an elected member of the European Faculty of Land Use and Development and his affiliations include the Association of Geographic Information, the European Union of Coastal Conservation and the Remote Sensing Society. He has published extensively in his areas of research interests.

Dr Robert Home is a Reader in Planning at the University of East London and holds degrees from Cambridge University, the London School of Economics and Oxford Brookes University. He teaches most aspects of town planning with emphasis upon developing countries. His main research areas are planning regulations, planning history and land management. He has published widely on planning practice and undertaken research and consultancy work in Nigeria, Malaysia, China, Bulgaria, Pakistan and South Africa.

Professor Crispus Kiamba is an Associate Professor in the Department of Land Development, University of Nairobi, Kenya. He holds a BA in Land Economics from the University of Nairobi, a MSc in Land Appraisal from the University of Reading and a PhD from the University of Cambridge. He is presently on leave from the Department of Land Development as the Deputy Vice-Chancellor in charge of Administration and Finance of the University of Nairobi. He has previously been the Principal of the College of Architecture and Engineering and the Dean of the Faculty of Architecture, Design and Development of the same University.

Professor Kiamba has main research interests in urban issues, land management and environmental and development issues. He is a Member of the Institution of Surveyors of Kenya. He has also been a Member of the Bureau of the International Federation for Housing and Planning and a Member of the Rural and Urban Planners in Southern and Eastern Africa.

Dr Reinfried Mansberger graduated from the Technical University in Vienna in Geodesy having majored in photogrammetry and Cartography.

His PhD thesis was on "developing a system for semi-automatic interpretation of tree crown conditions using colour infrared aerial photographs". Between 1983 and 1987, he was appointed as a university assistant at the Institute of Applied Geodesy and Photogrammetry at the Technical University at Graz before being appointed as a research assistant at the Institute of Surveying, Remote Sensing and Land Information, Universität für Bodenkultur Wien (University of Agricultural Sciences, Vienna). He is currently Assistant Professor at this institute.

Dr Mansberger is the Austrian correspondent of Commission VII (Resource and Environmental Monitoring) of the International Society of Photogrammetry and Remote Sensing. He is also actively involved in the International Federation of Surveyors as a correspondent member and webmaster of Commission 3 (Spatial Data Management) and is an elected member of the European Faculty of Land Use and Development. His principal research interests are: Land Use Planning, Land Information, Photogrammetry, Environmental GIS Applications, and Cadastral Systems.

Dr Walter Seher graduated in 1994 from the Univerität für Bodenkultur Wien (University of Agricultural Sciences, Vienna), in Rural Engineering and Water Management. He is currently appointed as a Lecturer in this university at the Institute of Regional Planning and Rural Development.

He has recently been involved in research projects concerning: the cultivation of alpine meadows and the implementation of Alpine National Parks in Austria; a comparative evaluation of three agri-environmental programmes in France, Germany and Austria; mobile flood protection elements; and the impacts of set-aside vineyards on cultivation, ecology, landscape image and the development of settlement structures in Lower Austria. His special fields of interest are: rural development, land consolidation, agri-environmental programmes, agriculture and nature conservation, and land use planning and water management (especially flood protection).

Robert Dixon-Gough and Reinfried Mansberger

The debate over agrarian structure in Macedonia: implications for land management

Dr. Peter BLOCH
The Land Tenure Center, Madison
Dr. Jolyne MELMED-SANJAK
Department of Economics, The University at Albany, Albany
Robert HANSON
Department of Agricultural Economics, The University of Wisconsin, Madison

Abstract

Macedonia entered the transitional period with a well-defined bimodal agricultural structure and the Ministry of Agriculture has, with the aid of the World Bank, proposed a programme. This paper describes a research project aimed towards establishing the relative productivity of the two agricultural structures with a view to recommending a strategy for land and agrarian reform. The paper describes the characteristics of the two types of farming activities and makes comparisons between the farms in the private and public sectors. This study concludes that despite the many disadvantages, the small farmers are producing profitably yet in spite of the many advantages, the social sector is not. It is therefore evident that the social sector of the Macedonian agricultural industry is in need of fundamental restructuring.

Introduction

Macedonia, like other constituent republics of the former Yugoslavia, entered the transition period with a bimodal agrarian structure. Seventy per cent of the agricultural land is owned by small, highly fragmented private farms averaging 2.5 hectares; most of the balance is held by vertically-integrated state agricultural enterprises (agro-kombinats) averaging 1,000 hectares. During the socialist period the two sub-sectors were characterised by a strong degree of symbiosis, but that has broken down with the dissolution of Yugoslavia and the drive toward privatisation.[1] It is clear to everyone that strong initiatives are required to redirect the agricultural sector to enable it to respond to the exigencies of a market economy. There is, however, a striking lack of consensus on what should be done.

Just like the agrarian structure, the policy debate about it is bimodal. The Ministry of Agriculture, which dealt only with the agro-kombinats during the socialist period, is convinced that the most serious problem to be resolved is the inefficiency of small private farms, which have scattered, tiny parcels and are frequently owned by 'part-time' farmers. The Ministry hopes to achieve greater economies of scale in the small private farm sector, via active programs for land consolidation. The World Bank, a major supporter of Macedonian development, is equally convinced that the biggest problem lies in the inefficiencies of the state sector, especially the excessively large landholdings of agro-kombinats. The Bank and other external donors are encouraging the privatisation and restructuring of the agro-kombinats, including the break-up of their big fields. This polarised policy debate, which has clear and fundamental importance for changes in land management, has been curiously unencumbered by facts until now.

In the first half of 1996, the Land Tenure Center (LTC) conducted in-depth research into the comparative productivity and profitability of farms in the state and private sectors. It also did a thorough assessment of the legislative framework that governs the agrarian sector. The findings are quite definitive:

- small private farms are quite productive and profitable in spite of the many institutional obstacles they face;
- agro-kombinats are not as productive as they should be in spite of the many institutional advantages they enjoy;
- fragmentation of the small farms does not have a generalised negative effect on productivity and profitability;
- the actual field size of the agro-kombinats is, on average, quite small (10-20 ha.) and thus amenable to cultivation by individual commercial farmers;
- there are substantial legal and institutional impediments to the operation of land markets.

The present paper explores the implications of the LTC research for the policy debate, and concludes that:

- active consolidation programs will not be cost-effective; land markets, if encouraged, should provide appropriate signals to develop a more rational agrarian structure;
- agro-kombinat land, especially that already divided into relatively small parcels, should be leased or sold to private farmers; the privatised successors of the agro-kombinats could participate in the competition to acquire the land, but on the same basis as other farmers.

Macedonian agriculture

The agricultural sector in Macedonia is bimodal, characterised by two very different farm enterprise types: small family farms operating on privately owned land and large, socially owned farms.[2] The socially owned farms can be further classified into two types:

1. agro-kombinats, vertically integrated agribusinesses which have very large landholdings and are engaged in primary production, extensive agro-industrial processing, commercial storage and marketing services; and
2. socially owned agricultural companies, which have smaller holdings and engage to a much lesser extent in non-primary production activities. The total arable land in Macedonia is 662,000 hectares of which 204,000, or about 30 percent, belong to socially owned farms. Most of the balance belongs to the private farm sector; the co-operative sector occupies a small percentage of the arable land.[3]

Many farm households derive a significant proportion of their incomes from non-farm activities; only 14 percent of the rural population of Macedonia is engaged in full-time farming by the official definition. The official definition is very strict: if any member of the household, not necessarily the head or spouse earns income from off-farm employment, the household is considered to be engaged in part-time farming. This definition is clearly inappropriate: it is very rare everywhere in the world that most farm families derive 100 per cent of their income from farming. Ongoing research on the data collected under this project will explore the possibility of alternative distinctions, including the definition of 'serious' farmers, those who are likely to respond to agricultural policy in a businesslike manner, as contrasted with 'hobby' farmers and subsistence farmers, who are less likely to do so.

There has also been substantial out-migration from regions where farm sizes are small or where agriculture is only marginally profitable. Furthermore, the population engaged in farming contains a high proportion of ageing and elderly persons. The family farm sector is composed of a wide range of farm operations, as Table 1 shows,[4] but most of them are small: the average farm size is currently estimated to be 2.5 - 2.8 hectares. There are some farms larger than 10 hectares, the legal maximum, because some families lease land.

Table 1 Landholdings of farm households

	Number	**%**
Total	*176,296*	*100.0*
Landless	2,354	1.3
> 1 ha.	78,735	44.6
1.01-2.00 ha	38,879	22.1
2.01-3.00 ha	22,170	12.6
3.01-5.00 ha	19,743	11.2
5.01-8.00 ha	9,380	5.3
< 8.01 ha	5,035	2.9

Source: Republic Statistics Office, Skopje. Census of population and households, 1981.

An important characteristic of the family farms is that they are fragmented: a family's landholding is not one parcel, but rather composed of several (sometimes more than twenty) non-contiguous parcels. In the 1981 Census, the average parcel size was 0.14 hectare; project survey results, shown in Table 2, indicate a larger average size. The fragmentation is generally due to inheritance practices and a long history of informal land market activity constrained by lack of capital and labour. A major concern of the Ministry of Agriculture is that this fragmentation causes production inefficiency and low output levels; the Ministry has suggested that an active government program to promote consolidation would be the best solution.

Table 2 Landholdings and family size of private farm sector

Region	Av. holdings size (ha.)	Av. number of parcels per holding	Average parcel size (ha.)	Av. number of family members p. household	Av. land per family member (ha.)
Mediterranean	2.30	8.3	0.277	4.42	0.52
Pelagonian	2.91	9.2	0.316	3.78	0.76
Western	1.79	5.7	0.314	6.04	0.30
Skopje/Kumanovo	2.65	5.5	0.482	4.84	0.55

Source: project survey results.

Another major problem facing family farms is the lack of both factor and product markets. Small farms are not able to take advantage of economies of scale in marketing. They have difficulty in obtaining inputs, no access to agricultural credit, no access to extension services or other information sources, few market outlets, and low prices for their products. The socially owned farm sector has acted as both factor and product markets for the family farm sector surrounding them. Since many of the agro-kombinats contained large processing plants, they purchased certain products such as wheat, vegetables and fruits from the family farms. The agro-kombinats have also provided family farms with necessary inputs and extension services. Many of these enterprises have greatly reduced their operations, however, because state subsidies have been cut and credit has been practically eliminated. Thus they are no longer able to provide the same level of services to the family farm sector, particularly at attractive prices or convenient terms. In addition, they are purchasing less of the family farm production and delaying payment for these purchases. While these tendencies reflect movement towards a more efficient agricultural sector, the development of alternative forms of providing such key services is very important during the transition to a private economy.

The large enterprises also absorbed surplus labour from the private farms in their areas. This important source of employment and wages for land-poor families will continue to shrink considerably with the restructuring of the agro-kombinats and other socially owned farms. While this reflects downsizing towards more efficient production strategies, in the immediate term it is leading to increasing unemployment, which has driven many towards the already constrained private small-scale production for subsistence and perhaps longer-run employment.

Privatisation of the agro-kombinats and socially owned farms is now inevitable, but the government is determined to maintain production levels in agriculture. The government has assumed that the productivity of the large-scale farming enterprises is considerably greater than that of the small-scale, private farming sector due to economies of scale, and therefore insists on maintaining the large fields that typify agro-kombinat agricultural production. One exception is the land that was expropriated from private owners since the late 1940s (by one estimate, this represents approximately 10 percent of the land used by socially owned enterprises). The previous owners or their heirs will be able to receive their land via a restitution process. Most of the land used by socially owned enterprises, however, has been under state ownership since before World War II and has been rehabilitated by the state through irrigation and drainage projects.

The land tenure and agricultural productivity project: agenda and method

Although the size, land tenure structure and type of production diverge significantly across the social and private farm sectors, there has been and continues to be a symbiotic relationship between the two sectors. The socially owned enterprises have acted as both factor and product markets for neighbouring family farms. Research conducted under this project found, however, that the closeness of this relationship varied widely, with some private farmers having no dealings at all with the agro-kombinats and others entering into secure marketing contracts with them. The transition period policy agenda, which is being defined for the two sectors, arises out of the fact that the government has significantly modified its policies and priorities with regard to these two types of farming enterprises. State policies previously supported production of the socially owned farm sector, neglecting the private farm sector to a large extent. Legislation and policies currently in development seek to make agricultural production a market-driven activity, abandon agricultural production by the state, and encourage the family farm sector to increase its productivity. This policy shift, together with general economic conditions and other macro-economic policy changes, means that both the private and socially owned farm enterprises are experiencing significant changes and facing new challenges. This project's objective was to conduct research whose results would enable the government to develop informed land tenure policy for a future in which the two sub-sectors will converge.

The Land Productivity Action Plan developed by the Ministry of Agriculture with assistance from the team of the Land Tenure and Agricultural Productivity Project outlined five tasks to be completed over the six-month term of the project:

- document and assess the land-related constraints to increased productivity and profitability of private farms;
- document and assess the land-related constraints to increased productivity and profitability of socially owned farms;
- assess the appropriateness of legislation, regulations and institutions affecting land tenure and land use;
- propose land policy adjustments that would promote the increased productivity and profitability of the agricultural sector;
- identify financial and technical assistance to support the development of land markets to promote efficient, sustainable and equitable increases in agricultural incomes.

In this paper, we focus on the first two of these tasks; an earlier version of our discussions of these tasks and the other three are contained in LTC (1996). The remainder of this section elaborates the relevant issues and delineates our data collection methodology. The next section of the paper presents the results for each sector and a comparison across sectors. The paper concludes with a policy discussion based on our data analysis and our assessment of the legal and institutional framework of agriculture in Macedonia.

Land-related constraints to increased productivity and profitability of private farms

With the transition to a market economy, the government must address several issues with regard to the private sector. At present, the principal constraints to increased productivity and profitability of private farming appear to be related to great uncertainty about market opportunities. This uncertainty is aggravated by the current decline and transformation of the socially owned enterprises and the inadequate development of appropriate alternative structures to provide services to the private farm sector.

As the government's agricultural policy begins to confront these immediate problems faced by small farms, it may be constrained by land tenure realities. As noted, farms are small and fragmented, and farming is frequently a part-time occupation of the landowners. There appear to be substantial constraints to the efficient use of land and labour resources due to excessive fragmentation: excessive amounts of cultivable land are wasted on border marking, possible crop damage due to incursion of tractors and persons accessing neighbouring plots. The market for agricultural land, other than for seasonal leases, is not very active and has historically contributed to fragmentation rather than yielding consolidation.

The research involved the selection of a sample of private farm households distributed across 4 of Macedonia's 6 agroclimatic regions.[5] A sample of 48 villages was selected from across these municipalities. The villages were stratified and chosen to include topographic variation (flat, hilly and mixed villages; high mountain villages were excluded as there is little crop cultivation in these areas) and to include a range of socio-economic conditions. A sample of 820 individual farm households was selected using stratified random sampling procedures. Stratification was according to farm size category and random selection of household was conducted from property registry books held by the municipal cadastral offices.

The farm size categories used are a simplification of the stratification scheme used by the National Statistical Office. Farms are grouped as follows: <1.00 ha., 1.01-2.00 ha., 2.01-5.00 ha. and >5.01 ha. Table 3 presents the population distribution of farms according to 1981 census data as well as the distribution of the sample. The number of the individual farms with land smaller than 1 ha. were reduced by 30% in the sample and that the number of farms larger than 2 ha. were increased to ensure adequate presence of larger farms in the sample to be able to address issues pertaining to size affects on productivity. Also, farmers without land are excluded from the sample.

Table 3 Population and sample farm size structure

	Number of Individual Farms		Percentage Distribution	
Holding size	*Population*[6]	*Sample*	*Population*	*Sample*
< 1.00 ha	78,735	265	45.3	32.7
1.01-2.00 ha	38,879	185	22.3	22.6
2.01-5.00 ha	41,913	288	24.1	35.1
> 5.01 ha	14,415	78	8.3	9.6
Total	173,942	820	100.0	100.0

Sources: Republic Statistic Office, Skopje. Census of population and households, 1981.

Two survey instruments were developed and administered to the President of each selected village and to the sample of households selected within each village. The questionnaires (village and farmer) contained open-ended opinion questions, multiple choice questions as well as several quantitative tables to be filled in during the interview. The village-level questionnaire was used to collect information about the village: population structure, production capacity, infrastructural development, economic activity of the rural population, levels of animal stock and machinery inventory and other general questions concerning the level of development of the village and the perceptions about the future of agriculture.

The farm household questionnaire consists of four major segments: demographic and social characteristics of the head of the household, spouse and other family members; land acquisition and land use history of the individual farm; agricultural production (including input and output data, capital stock and animal stocks); and agricultural marketing. In addition, a few open-ended, opinion questions were included to capture the farmer's individual perspective on the future of agriculture and the policy agenda.

Research was also conducted via detailed case studies in several villages, with the principal aim of gaining deeper insight into land transactions, costs and inheritance practices, and via a village-level survey to assess local variations in infrastructure, market access, and socio-demographic features.

Land-related constraints to increased productivity and profitability of socially owned farms

The average size of the arable landholdings of the socially owned farms is about 1,000 hectares, with considerable dispersion between the largest and the smallest (under 50 hectares to greater than 5,000 hectares). The current policy is to privatise the business operations of the agro-kombinats and socially owned agricultural companies, but to retain the agricultural land in state ownership; the land would be leased according to, as yet, undetermined procedures to private farmers or the successor enterprises of the agro-kombinats. The government is reluctant to break up the large fields into smaller units suitable for cultivation by individual farmers because it fears the loss of economies of scale as well as the same process of fragmentation that has occurred in the private sector.

The project studied the land use, productivity and profitability of farming on the land currently under the control of the agro-kombinats. Field level data were collected from a sample of ten farms and were combined, as far as possible, with disaggregated input and output information in both physical and financial terms. In addition, the research team assembled documentation on world wide experiences with leasing of publicly owned agricultural land.

Ten agro-kombinats were selected to include varying agronomic, climatic, organisational and economic conditions. Particularly significant factors in the choice of enterprises which we analysed were the size of the agricultural and arable area land holdings, the farm's location, crop patterns, as well as the status of the records maintained within the enterprises.

Because access to social sector data is difficult and because of the complex and uncertain situation surrounding these enterprises currently, we chose to follow a case-study approach for a representative group of enterprises rather than using random sampling.[7]

Based on these criteria, we selected ten agricultural enterprises which are stratified into three size categories: 3 small enterprises (with land area up to 1000 hectares); 4 medium enterprises (with land area of 1000 - 5000 hectares); and 3 large enterprises (with land area over 5000 hectares). The location of the enterprises by agro-climatic region was also taken into consideration while making the choice of the representative sample.

Each enterprise was visited on multiple occasions. The first visit set the stage for implementation of the survey and to gather general enterprise descriptions. On the second visit, a detailed questionnaire regarding the enterprises' land acquisition history and its 1995 land use and production activities was delivered. A person from the farm's technical managerial staff was selected to provide the responses to the questions. Further visits for review and assistance in completion of the questionnaires were conducted as needed according to the case.

Research Findings

Comparative productivity and profitability of small farms and agro-kombinats[8]

Table 4 summarises the information on yields gathered from the surveys of private farmers and agro-kombinats. For some crops, notably wheat, which is by far the most prevalent crop in Macedonia, the productivity of small private farms, as measured by output per hectare, is greater than that of the agro-kombinats. For others, such as tomatoes and alfalfa, the agro-kombinats achieve significantly greater yields. Yields of barley and wine grapes are comparable. The two crops for which the agro-kombinats outperform the private sector are special cases, because production technologies are very different in the two sectors. Agro-kombinats produce most of their tomatoes in greenhouses, which are rare among private farmers. Similarly, much of the alfalfa is grown under irrigated conditions in the agro-kombinats but without water control or with poorer irrigation infrastructure on private farms.

Table 4 Output per hectare, 1995 (kg./ha., weighted average)

CROP	Mediter-ranean	Pelagonian	Western	Skopje-Kumanovo	Agro-Kombinats
Wheat	3153 n=200	3412 n=109	3176 n=90	3155 n=141	2463 n=10
Corn	4139 n=125	5604 n=26	3554 n=108	3812 n=104	na
Barley	2538 n=119	3138 n=48	0 n=0	3079 n=75	3135 n=7
Rice	5843 n=14	0 n=0	0 n=0	0 n=0	na
Tomatoes	29159 n=96	22111 n=30	2505 n=10	10838 n=13	97541 n=5
Wine Grapes	10571 n=108	0 n=0	0 n=0	0 n=0	9170* n=8
Alfalfa	5611 n=48	4374 n=13	4661 n=42	5548 n=33	11778* n=5

The results for wheat and barley were contrary to the conventionally held opinions of academics and agriculture officials in Macedonia; those opinions were formed in the absence of prior systematic research. The government's determination to maintain the large-scale production of grain on land now held by the agro-kombinats was based to a great extent on its assumption that there were economies of scale which would be lost if the fields were subdivided. Our research shows, however, that small-scale production can produce similar or greater output at lower costs per hectare.

To achieve a greater understanding of the underlying reasons for these productivity and cost differences, it is useful to examine the intensity of input use and values. Table 5 shows the physical quantities of inputs for wheat production in the private and agro-kombinat sectors; we further identify the results for the smaller agro-kombinats, those whose scale is closest to that of private farms. With the exception of the substantially greater inputs of labour and machine hours in the Western region (where farms can be said to be both capital-intensive and labour-intensive!), input intensities do not appear to be substantially different in the two sectors. Small agro-kombinats are more input-intensive than all agro-kombinats, and achieve a higher than average yield. But they still achieve considerably poorer productivity than does the average private farmer.

Table 5 Input use and output per hectare in wheat production (weighted averages)

	Pelagonia	Western	Skopje/ Kumanovo	Mediter- ranean	Agro- kombinats	
Inputs (per ha)	n=103	n=96	n=112	n=106	all n=10	small n=3
Seed (kg)	280.8	360.2	295.1	294.3	283.8	284.3
Fertilizer (kg)	335.8	369.5	354.4	274.0	278.4	477.9
Pesticide (litres)	1.9	3.9	4.2	1.4	5.2	1.7
Labor (hours)	34.2	181.6	46.7	52.4	39.8	61.7
Machine (hours)	23.9	33.4	20.9	22.8	14.7	22
Yield	3189	3318	3041	2902	2463	2789

The same results hold in value terms, as Table 6 shows: private farmers achieve greater revenues per hectare at a lower input cost per hectare than the agro-kombinats. They spend more on fertiliser, and in three of the four regions more on pesticides, but their total costs per hectare are about half as great as those of the agro-kombinats. Even the addition of the imputed cost of family labour does not eliminate the profitability of private farms, although in the Western region, profitability drops considerably because of the great number of hours per hectare applied to wheat in that region.

Table 6 Value of inputs and output per hectare of wheat production (weighted averages)

Value of Inputs (D/Ha)	Pelagonian	Western	Skopje/ Kumanovo	Mediter- ranean	Social Sector Sample
Seed	4903	7965	5817	4863	5826
Fertiliser	3651	3942	4073	4488	2654
Pesticides	666	1878	2339	2349	1589
Total Cost/ha[9]	14327 15995[10]	18128 27934	15440 17926.0	13954 16690	29432
Total Revenue/ha	32464	33777	30957	29542	28910
Net Revenue/ha	19108 17478	15496 5442	15566 13005	15585 12849	-522

Wheat was not the only crop for which agro-kombinats did not produce profitably. Table shows that they only achieved positive net revenue for alfalfa and fodder peas, both intermediate goods whose profitability may have been artificially determined by high internal transfer prices, and for apples.[11] By contrast, private farmers achieved profitability on many crops, even when family labour was valued at the market wage. A major explanation for the unprofitability of agro-kombinat production was the high level of overhead costs (management expenses, interest on debt, etc.), that are not incurred by small private farms.[12] In other words, diseconomies of management are much more important in Macedonia than scale economies.

It must be recognised that 1995 was a particularly difficult year for Macedonia, in general, and in particular for its agricultural sector. Borders were closed with Greece on the south and Serbia on the north, and there have never been particularly close relations with Albania on the west and Bulgaria on the east. Thus there was a substantial drop in Macedonia's exports, a large percentage of which are agricultural, and a concomitant drop in its ability to import agricultural inputs. Furthermore, macroeconomic changes were also unfavourable for agriculture, with high inflation and increasing unemployment due to the shutdown of many state-owned industrial and commercial enterprises.

The government suggests that these factors were the principal contributors to the poor performance of the agro-kombinats in 1995. This is probably true. But it is also true that the same factors affected the private farm sector, in precisely the same manner. Thus the entire agricultural sector was hurt by exogenous phenomena, but private farmers appear to have been better at coping with them.

Effects of fragmentation on productivity and profitability of small private farms

Fragmentation of private farmers' landholdings is extreme in Macedonia, as Tables 8 to 11 demonstrates. In each region, farms have an average of five to eight parcels; five per cent of them have 16 or more. Parcels average less than one-half hectare in size, and are scattered over a wide area, with most farmers having to travel three or more kilometres to the farthest parcel. While average parcel size increases with farm size, so does the number of parcels; thus the Janusewski fragmentation index[13] is lower (i.e. fragmentation is more extreme) for larger holdings, suggesting that an increase in the size of holdings alone will not solve the fragmentation problem. This statistical picture confirms the impression held by Macedonian officials and academics whom are concerned that fragmentation constitutes a major constraint to the productive potential of private farms.

Table 7 Cost structure of selected products and profitability of production on agro-kombinats: in percent

Indicator	Wheat	Barley	Sun-flower	Fodder Peas	Alfalfa	Toma-toes	Grapes	Apples
Seeds	19.79	13.83	5.07	36.92	2.89	3.80		
Fertiliser	9.06	8.12	8.06	3.61	2.31	2.80	1.86	2.61
Protective measures	5.40	4.25	8.76	0.00	0.45	2.79	14.58	13.37
Mechanical serv.	25.72	25.35	25.68	25.47	27.80	2.79	6.52	10.00
Deprecia-tion	1.74	0.89	0.81	3.75	2.30	13.79	7.94	13.13
Salaries & wages	11.28	9.96	19.28	4.98	13.72	24.35	26.33	20.27
Subtotal "direct"	72.99	62.4	67.66	74.73	49.47	50.32	57.23	59.38
Insurance	3.27	4.22	5.43	0.42	0.29	3.52	9.36	11.26
Manage-ment costs	8.18	8.32	7.14	12.45	26.87	17.93	24.84	6.98
Interest costs	8.73	11.90	5.51	4.05	2.59	0.00	0.48	2.04
Costs of sales	1.85	1.45	1.85	1.77	0.28	0.00	0.35	10.07
Other costs	4.98	11.71	12.41	6.58	20.50	28.23	7.74	10.27
Subtotal "indirect"	27.01	37.6	32.34	25.27	50.53	49.68	42.77	40.62
Total	100.0	100.0	100.0	100.0	100.0	100.0	100.0	100.0
Profit (%)	-1.77	-11.38	-10.49	5.52	103.48	-2.34	-10.64	2.08

Simple correlation coefficients were estimated between the fragmentation index and yield, per hectare costs and per hectare profits in wheat production. In the Pelagonian and Mediterranean regions, no significant correlation was observed. In the Western region, the index of fragmentation is negatively and significantly correlated with wheat yields and per hectare production costs. In the Skopje-Kumanovo region, the index correlates significantly but positively with costs of production and negatively with profits. This could reflect the combination of the positive relation between farm size and index on one hand and a possible qualitative change in the package of inputs used in the larger farms. Overall, there is not a systematic significant relationship between fragmentation and the productivity of small farms. In other words, fragmentation does not appear to be a generalised problem. In certain regions and for certain crops, however, productivity might be increased if parcels were larger.

Table 8 Degree of land fragmentation by farm operation size and by region - Mediterranean

Size structure of farms (ha)	Number of farms in sample	Mean plot size (ha).	Mean no. of plots per farm	Mean of Janusewski's Fragmentation Index	Mean distance (km) to farthest parcel	Mean distance (km) to nearest parcel	Believe land is too fragmented (%)
< 1.	85	0.214	3.76	.53	2.8	.98	89
1.01-2	81	0.274	6.80	.45	3.05	.92	91
2.01-5	112	0.390	11.48	.35	3.14	.60	90
> 5.	28	0.629	13.17	.36	4.03	.44	93
Total*	306	0.305	7.16	.45	3.05	.883	90.1

Table 9 Degree of land fragmentation by farm operation size and by region – Pelagonian

Size structure of farms (ha)	Number of farms in sample	Mean plot size (ha).	Mean no. of plots per farm	Mean of Janusewski's Fragmenta-tion Index	Mean distance (km) to farthest parcel	Mean distance (km) to nearest parcel	Believe land is too fragmented (%)
< 1.	23	0.247	3.00	.63	1.75	.52	68
1.01-2	39	0.321	5.07	.48	2.06	.40	72
2.01-5	60	0.328	11.35	.33	2.52	.45	83
> 5 ha.	26	0.519	16.32	.30	3.06	.53	85
Total*	148	0.329	7.84	.46	2.24	.47	78.2

Table 10 Degree of land fragmentation by farm operation size and by region – Western

Size structure of farms (ha)	Number of farms in sample	Mean plot size (ha).	Mean no. of plots per farm	Mean of Janusewski's Fragmen-tation Index	Mean distance (km) to farthest parcel	Mean distance (km) to nearest parcel	Believe land is too fragmented (%)
< 1	31	0.217	3.48	.55	2.21	.91	76
1.01-2	52	0.273	5.60	.46	3.00	.74	88
2.01-5	43	0.456	7.12	.43	3.24	.85	86
> 5	5	1.030	8.40	.42	2.90	.27	
Total*	131	0.328	5.36	.48	2.79	.80	83.2

Table 11 Degree of land fragmentation by farm operation size and by region - Skopje-Kumanovo

Size structure of farms (ha)	Number of farms in sample	Mean plot size (ha).	Mean no. Of plots per farm	Mean of Janusewski's Fragmentation Index	Mean distance (km) to farthest parcel	Mean distance (km) to nearest parcel	Believe land is too fragmented (%)
< 1	32	0.238	3.66	.59	1.73	.65	85
1.01-2	50	0.374	4.92	.53	2.78	.88	86
2.01-5	62	0.777	6.66	.48	2.64	.53	77
> 5	24	1.281	6.58	.45	4.92	.72	61
Total*	168	0.626	5.55	.51	2.83	.69	79.2

On the other hand, many farmers consider fragmentation to be a problem and express willingness to participate in consolidation programs (Table 12). As in other aspects of their ideas about farm policy, farmers feel that the government should play a leading role in promoting consolidation.

Yet, in spite of their desire for government programs, farmers are behaving in a way that reduces the fragmentation problem: they are purchasing and leasing land, thereby increasing their operational holdings sizes. As Table 13 shows, the purchased and leased parcels are larger than the parcels acquired via inheritance. Also, without exception, the average size of purchased parcels is largest amongst the group of farms operating with over 5 hectares. Under current law, there is an upper size limit of 10 hectares of land held in ownership; some farmers exceed that limit through leasing.

Table 12 The problem of fragmentation and what to do about it

	Is fragmentation a problem ?		Should government programs promote consolidation ?	
Regions	YES	No	YES	NO
Mediterranean	90.2%	9.8%	85.3%	14.7%
Pelagonian	78.1	21.9	73.6	26.4
Western	82.8	17.2	66.9	33.1
Skopje-Kumanovo	79.2	20.8	67.9	32.1

Source: project survey data

Considering the government's reluctance to consider the break-up of the landholdings of the agro-kombinats, it is interesting that the holdings of the state enterprises are fragmented, too. The average parcel size in most crops is surprisingly small: about 10 hectares and the average parcel is bigger than the total holdings of nearly all private farmers, but it is not of a size that makes it inconceivable that private farmers could assume its management without subdivision.

Land market operation

The data regarding the purchase and lease of agricultural land in the four sampled regions in Macedonia suggest that a land market does exist, but that its activity is unstable. Purchases have declined sharply over the past several years and long-term leasing is rare. This is due at least in part to increased perceptions of risk due to macroeconomic, sectoral and political factors: inflation, overvalued exchange rates, removal or revision of subsidies, closed national borders. A return to higher levels of transactions will await the resolution of macroeconomic and sectoral problems.

Table 13 Land acquisition strategies by farm size

Farm Size in ha	Number of Farms[14]	% of Farm Inherited	Mean Inherited Plot size	% of Farm Purchased	Mean Purchased Plot size	% of Farm Leased	Mean Leased Plot size
A. MEDITERRANEAN							
< 1	85	(64%)	0.166	(18%)	0.199	(16%)	0.291
1.0-2.0	81	(78%)	0.210	(10%)	0.244	(6%)	0.316
2.01-5.0	112	(73%)	0.257	(13%)	0.320	(4%)	0.318
> 5	28	(67%)	0.541	(22%)	0.339	(6%)	0.604
Total[15]	306	70%	0.229	15%	0.250 ha	6%	0.351
B. PELAGONIAN							
< 1	23	(66%)	0.196	(15%)	0.164	(15%)	0.383
1.0-2.0	39	(84%)	0.255	(9%)	0.397	(7%)	0.417
2.01-5.0	60	(89%)	0.280	(9%)	0.329	(1%)	0.326
> 5	26	(77%)	0.385	(9%)	0.774	(14%)	0.907
TOTAL	148	78%	0.262	11%	0.381 ha	9%	0.455
C. WESTERN[16]							
< 1	31	(72%)	0.162	(24%)	0.214	(4%)	0.165
1.0-2.0	52	(84%)	0.224	(11%)	0.240	(5%)	0.458
2.01-5.0	43	(75%)	0.369	(19%)	0.509	(2%)	0.290
> 5	5	(74%)	0.677	(21%)	0.667	(4%)	0.600
TOTAL	131	77%	0.257	18%	0.313	4%	0.325

Farm Size in ha	Number of Farms[17]	% of Farm Inherited	Mean Inherited Plot size	% of Farm Purchased	Mean Purchased Plot size	% of Farm Leased	Mean Leased Plot size
D. SKOPJE-KUMANOVO REGION							
< 1	32	(94%)	0.173	(5%)		(1%)	
1.0-2.0	50	(87%)	0.252	(10%)	0.468	(3%)	0.383
2.01-5.0	62	(92%)	0.456	(5%)	0.710	(1%)	0.490
>	24	(71%)	0.818	(21%)	1.530	(8%)	1.444
TOTAL	168	89%	0.356	8%	0.744	3%	0.599

Source: project survey data.

There are also significant legal impediments. Pending private claims to socially owned agricultural land, lack of credit, and difficult political and marketing conditions are all possible contributing factors to the low level of land transactions. Current legislation and legal processes also make the sale and purchase of land somewhat complex. Transaction costs, as a result, are reported to be quite high as the procedures for selling and purchasing land usually require a lawyer and involve approval from numerous agents. According to existing regulations, an individual interested in selling land must first, usually through a lawyer, obtain the following documentation:

- contract for purchase to register the contract;
- certificate from the co-operative and/or agro-kombinat in the cadastral area stating that it is not interested in purchasing the land;
- statements from the owners of neighbouring plots, stating that they are not interested in purchasing the land;
- certificate from the Municipal office of the Ministry of Civil Engineering, stating that the land is not part of construction land (i.e. land zoned for future residential and commercial construction);

- certificate from the Municipal office of the Ministry of Finance stating that the land has not been nationalised.

Not only do these legal steps take money and time - up to several months, they also act as serious constraints because of the possibility, however slight, that some of the individuals and institutions whose consent is needed might have reasons for refusing to give it.

Conclusion

Despite the above caveats, a fair research conclusion is that despite many disadvantages, the small farmers are producing profitably; and, in spite of many advantages, the social sector is not. The data and analysis presented here challenge some prevailing stereotypes. For example, it is said that the social sector farms can not be privatised nor their land restructured for several reasons: first, because the social sector is, and has always been, more productive - from a smaller land base, 20% of the nation's arable land, 50% of the nation's agricultural product is generated; and second, because the social sector farms produce at a scale and level of technology which small farmers can not manage. Third, the small farmers are backward and unproductive. Our data have challenged these stereotypical images and instead presented a much more varied picture. For example, our report dampens the impact of the statement about the relative productivity based on the percent of land/percent of output comparison. We have noted the following qualifications to such comparison: the concentration of marketing capacity in the social sector, the favourable access to inputs that the social sector farms historically had, the relatively good private sector yields for certain crops, and the discrepancy in how much land is under the management of private farmers as reported in the census and the amount of land the official statistics includes under the category of private sector. Next, while there are certain clear achievements in the social sector which might not have been accomplished from the pre-socialist private farm sector (such as investments in extensive greenhouses and processing facilities), the private sector is beginning to demonstrate an ability to enter the processing side of agriculture e.g. some private dairies doing well, a private rice mill in Kocani and private input suppliers are emerging. Also, if we pointed to the fact that the average parcel size farmed in the social sector is not beyond the scope of individual farmer or small, private agricultural enterprise management. Finally, we provide some evidence of productivity in the sector of small, private farms.

Our conclusion is, therefore, that the agricultural sector is in need of fundamental restructuring so that the mismatch in allocation of land and labour resources: the under utilisation of land in social sector, the over skilled and over sized workforce in the social sector combined with the converse conditions in the private farm economy farm sector, is eliminated. As long as an appropriate framework for restructuring is created and the legal and institutional bases of a market economy are developed, restructuring should yield a rationalisation of enterprise and production strategies. This will occur as incentive-based decisions lead to eliminating diseconomies of scope which characterise the huge, complex social sector farm enterprises, and to the guiding of land to its most productive uses and users. These uses and users will be diverse and will include in some instances a continuation of the present management of land and in other instances will entail reallocation.

References

Burton, S., & King, R., 1982. Land fragmentation: notes on a fundamental rural spatial problem, *Progress in Human Geography,* **6**(4), 475-495.

LTC, 1996. *Socio-Economic Analysis of Agriculture in the Republic of Macedonia,* Draft report to the US Agency for International Development, prepared by the Land Tenure Center, The University of Wisconsin, Madison, WI, July 1996.

Januszewski, J., 1968. Index of land consolidation as a criterion of the degree of concentration, *Geographia Polonica,* **14**, 291-296.

Republic Statistics Office, 1981. *Census of Population and Households,* Skopje, Macedonia.

Endnotes

1. Our research found evidence that not only has this relationship deteriorated, but that even during the socialist era the nature and extent of symbiosis between agro-kombinats and neighbouring communities varied across villages and across ethnicities. In some instances the relationship resembled contract farming with its variety of input, credit and output market linkages; in other cases the relationship seems to have been more truly co-operative and community development-spirited; and in still other cases there was no

relationship to speak of.

2. Social ownership is a concept used in the former Yugoslavia to describe assets 'owned by everyone and by no one'. Different in concept but not in practice from state ownership.

3. There are serious discrepancies among data sources even for such fundamental numbers as the cultivated area. The 1994 Census reports that private farms cultivate about half the amount that the Statistics Office reports. There is also imperfect reporting of the subdivisions within the social sector, among the organised social sector (agro-kombinats), the unorganised social sector (scattered parcels acquired by the state over time) and the co-operative sector.

4. The 1981 Census, the source of these data, was the last published source of national information on land distribution. The 1994 Census is in the data entry process; it will certainly show a decline in the average size of farm holdings.

5. The four regions studied were Western, Skopje-Kumanovo, Pelegonia, and Mediterranean. The Big Lakes and Eastern regions were considered less likely to contain farms with the potential for productivity increases in response to policy reforms.

6. These data are taken from the Census of Population in 1981. According to the census, farms are defined as follows: a) every family with land holdings of at least 10 square *are* which can be used for agriculture; or, b) a family with land smaller than 10 square *are* if it has at least: 1 cow and calf or sheep and lamb, or, 1 cow and 2 grown head of small cattle, or, 5 grown sheep, or, 5 grown pigs, or, 4 grown sheep and pigs together, or, 50 head of grown poultry, or, 20 bee hives. Agricultural land holdings of individuals consist of land that is owned by the members of the family as well as the land that is owned by other entities, which was used by the family during the time of the census.

7. It should be noted that a factor for selection of the enterprises was also the existence of prior contact with the enterprise via prior research or otherwise. Despite this fact, the data were objectively reported to the extent that objective data existed in the farm's records.

8. There is no evidence of systematic bias in reporting by private farms as compared to social-sector farms. If there were such a bias, however, the Macedonian members of the research team suggest that private farmers would tend to underestimate their yields and social-sector farms to overestimate theirs.

9. Includes other costs, such as labour, machine hours, depreciation, and various management costs.

10. Italicised values include an imputed value for non-remunerated family labour which values family labour at an average/approximate market wage rate which such labour could earn working on another private farm in the Skopje/Kumanovo area; it is consistent with the wage rate implied for field labour in our sample social enterprises. In future efforts, it will be necessary to use more precise, region-specific wage rates.

11. Evidence for this is that lamb meat and cow's milk, for the production of which alfalfa and peas are important intermediate goods, are among the least profitable outputs of the agro-kombinats.

12. The argument could be made that some of these costs should be imputed for private farms for the same reason as family labour: such items as insurance and interest.

13. This index (K), with (a) representing parcel size, is defined as:

$$K = \frac{\sqrt{\sum a}}{\sum \sqrt{a}}$$

The index was developed by Januszewski (1964). It divides the square root of the total farm area by the sum of the square roots of the plot sizes. This index ranges between 0 and 1, with a value of 1 indicating a farm operation with one contiguous parcel. The index has three properties; fragmentation increases (the value of the index decreases) as the number of plots increases, fragmentation increases when the range of plot sizes is small, and fragmentation decreases when the area of large plots increases, and that of small plots decreases (Burton & King, 1982).

14. In the Mediterranean Region, the acquisition of land by 'gift', also accounts for 5% of the land acquired across all size categories. (All percentages do not total 100% as land acquired by gift is not included in the tables.

15. All totals are weighted averages according to the population distribution across farm sizes.

16. In the Western region, land acquired by 'gift' also accounts for 2% of all land acquired.

17. In the Mediterranean Region, the acquisition of land by 'gift', also accounts for 5% of the land acquired across all size categories. (All percentages do not total 100% as land acquired by gift is not included in the tables.

Implementation of land reform in Estonia

Prof. Dr. Kari I. LEVÄINEN
Helsinki University of Technology

Abstract

This paper identifies and discusses the three programmes of land reform, which have been undertaken during this century, although the emphasis is placed upon the land reform programme currently being implemented as the result of the 1991 Acts on the Principles of Property Reform and Land Reform. The general objectives of these Acts are to restitute the rights of expropriated land as the economic and social basis of private market economy.

The processes of land reform are examined and the role of land registration and the cadastre in the process of land reform are described. The current land reform in Estonia combines both the restitution of land and the formation of real property. For this purpose, the National Land Board requires information concerning the owners, cadastre, values and land use data. The processes by which this data is gathered for land restitution and formation are described and commented upon.

Finally, the paper outlines some of the problems that have been identified, including the lack of resources and adequate staff training.

This paper is an updated version of an article published in "Surveying Science in Finland".

Land reforms in Estonia

Estonia has undergone three land reforms in this century. The first Republic carried out an agrarian reform was executed in 1918. The second reform took place in the 1940 s, and the third is currently underway (Vallner 1993a.). When Estonia gained independence in 1918, 58% of the farmland belonged to large estates, mostly owned by Germans. The first Republic eliminated these estates and created about 50,000 new private farms.

The early Soviet rule commenced by nationalising land and most existing buildings. After 1940, farms of up to thirty hectares could be farmed privately (Wulff *et al* 1994). The collectivisation was completed later in the 1940s (Raitanen & Tenkanen, 1992).

The third land reform began already before the end of Soviet rule in Estonia with the Farm Act (Dec. 8, 1989). This Act was very significant to the development of Estonian agricultural policy, because it provided alternatives to kolkhozes and state farms. The area of these farms averaged twenty two hectares. Around 8300 farms were formed (Vastel, 1993) through this Act.

Today the land reform continues and the re-establishment of the cadastre is underway. The Act on Property Reform Principles (June 13, 1991) came into force on June 20, 1991. For example, this Act provides, as a rule, that unlawfully expropriated land should be returned to the previous owners. It contains principles concerning the legal subjects of the property reform and defines types of properties to be restituted. The Land Reform Act (Oct. 17, 1991) came into force on Nov. 1, 1991.

The general objectives of the National Land Board in the ongoing land reform are the following (Vallner 1993; Wulff *op cit*):

- to restitute the rights to private property and the status of individual ownership of property as the economic and social basis for a private market economy;
- to privatise land still owned by the state and not to be restituted; this land can be used for replacement or transferred to people having a preferential right (replacement was repealed in 1996) while avoiding new injustice, to restitute the rights of persons who were previously caused damage or to compensate such damage;
- to reserve state property for administrative and other special and defined functions such as defence, research studies, training, development etc.;
- to increase the influence of the municipalities on land management.

The land reform allows old landowners to (Leväinen 1993):

1. get back their land at the same place *(restitution)*

2. receive compensation as land at a new place *(replacement)*
3. receive compensation in bonds or money *(compensation)*
4. give up their rights to land entirely or to the advantage of someone else *(abandon)*.

If no application is submitted for a property, it changes into reserve land. The replacement of land at another place of an equal value was also possible up to 1996. But replacement did not function in practice. That is why the statute was repealed in 1996. For the compensation procedure, there are bonds that can be used for buying land, shares or flats (Vatsel 1993). The Estonian Government makes the decisions concerning state and municipal land. Bonds are nowadays freely transferable, the market value is about 16-18 % of their face value.

State farms and kolkhozes have been abolished. The Liquidation Commission for buildings and structures in state farms and kolkhozes has nearly finished its work. There are sub-commissions in every municipality and kolkhoz. Roads, as well as planned road areas, remain as the property of state or municipality. Maintenance is the responsibility of the state or municipality.

According to the agricultural reform it is possible to release land, which has been used by state or collective farms, private companies or state enterprises, for agricultural production for a period of one to three years. This kind of holding will be changed into real ownership.

Land reform process

The implementation of the land reform includes many technical and legal procedures (Figure 1) which can be grouped as follows (Leväinen *op cit*):

- restitution and compensation
- real property formation
- registration
- leasing and selling of state land
- mapping
- assessment.

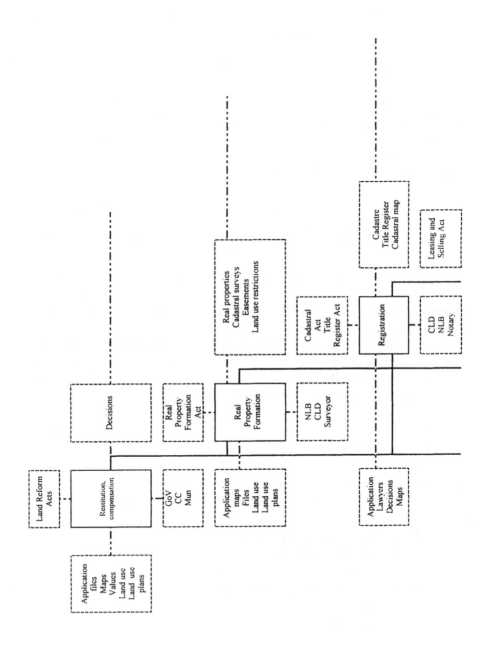

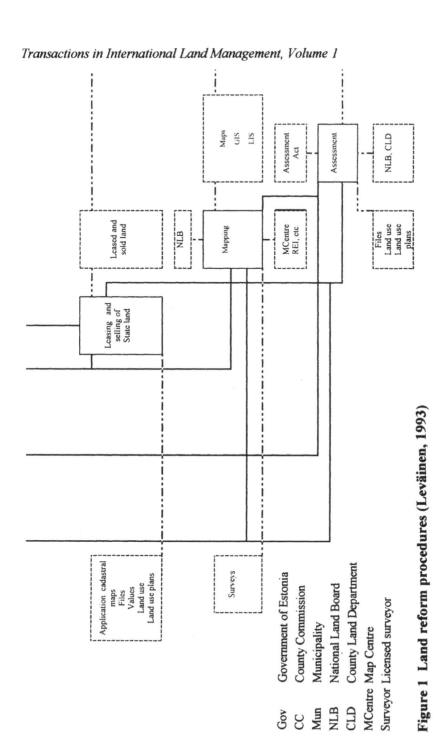

Figure 1 Land reform procedures (Leväinen, 1993)

Gov   Government of Estonia
CC   County Commission
Mun   Municipality
NLB   National Land Board
CLD   County Land Department
MCentre Map Centre
Surveyor Licensed surveyor

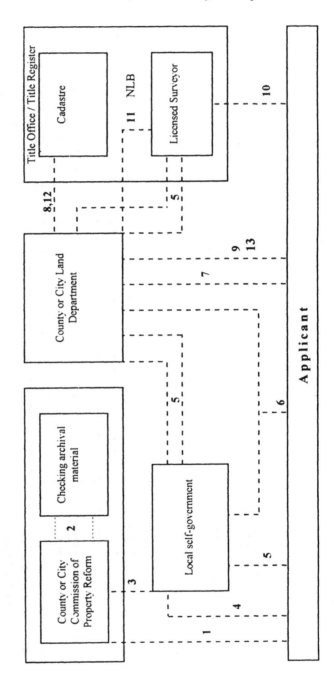

1. Application for getting land restitution
2. Checking the archives material, definition of the subject who has the right to the land entered into the register
3. Decision of the County Commission about the verified subject (the legal owner)
4. Definition of terms for land restitution
5. Establishment of real property and co-ordination with the file, working out a real property plan (a project to restitute the land parcel)
6. Decision of municipality of restitution of land
7. Application for registration of ownership rights
8. Registration of right of ownership of real property and land
9. The map of real property (the project of restitution of land together with an extract from the land cadastre)

In case there is no established boundaries:

10. Application for demarcation of real property with permanent boundaries
11. The map of real property
12. Registration of permanent boundaries and the map of real property with measurements
13. A map of registered real property

Figure 2 Restitution procedures and formation of real property (adapted from Maareformi täna, 1993)

The Government, the County Commissions and the municipalities are responsible for making decisions on land restitution and compensation. The formation of real properties is a technical process that includes checking of documents, cadastral surveys and the preparation of documents and maps. The National Land Board, the County Land Departments and licensed surveyors carry out these tasks. Land registration is organised in two parts: the National Land Board assisted by the County Land Departments is responsible for the management of the cadastre; the Ministry of Justice assisted by County Courts is responsible for the title register. Title Offices and title books were re-established in 1994. The parliament has accepted the Title Register Act in 1993 and the Cadastre Act in 1994.

The County Governor can make lease contracts for 50 years on state-owned land and will later also sell such land. The Government can lease land for 99 years. During the land restitution process it is possible to get permission for temporary usage for three years.

Land values are needed both for the execution of the land reform and for taxation. Real properties have been assessed in 1993 and 1996 for taxation purposes. The Assessment Act came into force in 1994.

Organisation of restitution and compensation procedures

The National Land Board, the County Land Departments and Commissions, and the municipalities are involved in the decision-making concerning the land reform. The National Land Board has prepared the preceding chart (Figure 2) of the process that includes both the restitution of ownership and real property formation.

January 17th, 1992, was the first deadline for applications to the local commission of ownership reform. Then an additional time was given for submitting restitution applications. The final deadline was March 1993. On the basis of documents filed in archives (cadastral and ownership data and cadastral map 1939-40), the County Commission will investigate whether the relevant property has belonged to the applicant, the applicant's parents or other relatives, and decide whether to declare the applicant the legal owner. The municipality where the property is situated will be notified of the decision. 215,000 applications had been submitted for restitution (Vallner *op cit*). For instance, in the city of Tallinn alone, 10,000 applications have been handed in.

The municipal land office, in co-operation with the applicant, will decide on such practical questions as when and how the property should be restituted and the boundaries marked. The documents required for a restitution decision will be produced by the municipal land office or ordered from the County Land Department.

On the basis of these documents the County Land Department will open a restitution file on the land parcel (formation of real property) and prepare a schedule for its restitution process. In practice the County Land Department orders this work from licensed surveyors, who submit all documents to the County Land Department for supervision. Field surveys are carried out at this stage. The relevant documents will finally be sent to the municipality in question.

Up to this stage it is possible for the applicant to withdraw or change his/her application for restitution, compensation etc. Otherwise the municipality will proceed to decide whether to restitute the land parcel, to determine the land use restrictions and to deliver the documents to the County Land Department for the registration of the relevant property in the state cadastre.

After the decision of the municipality has been confirmed, the legal owner will have to apply to the County Land Department for the registration of the real property in the state cadastre. The County Land Department forwards the application to licensed surveyors, who return the documents to the County Land Department.

After the registration of the real property, the legal owner will receive an extract from the state cadastre, a real property map and other documents. These are temporary certificates of the real property restitution. A temporary certificate will entitle the holder to use and lease the land parcel but not to full ownership. Before full ownership is confirmed, the holder is not entitled to sell, bequeath or give away the land parcel. Furthermore, the borders of the real estate should be marked with permanent border marks and surveyed by the County Land Department. This system of temporary certificates was abolished in 1993.

After the parcel of land has been registered in the state land cadastre, the legal owner is entitled to apply for the registration of his ownership in the title register.

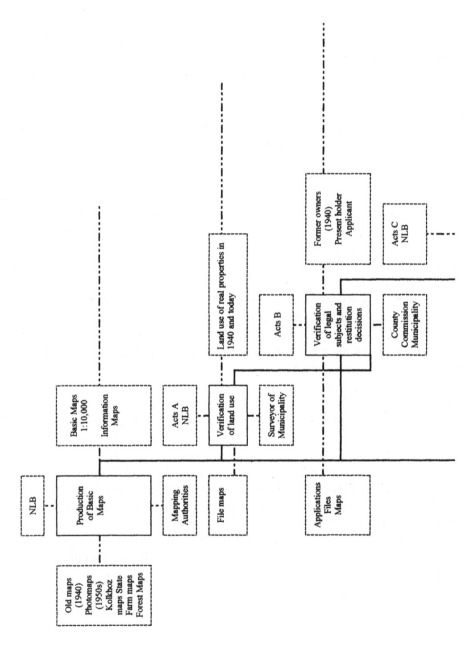

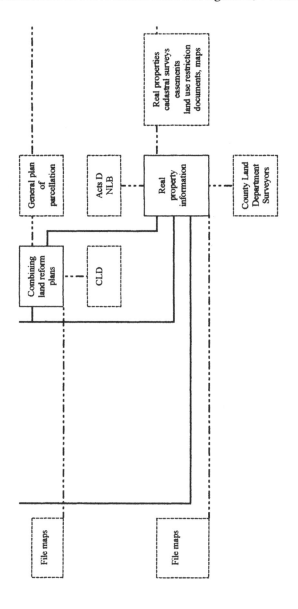

Figure 3 Stages of land restitution and formation (after Leväinen, 1963)

Real property formation

The current land reform in Estonia combines the restitution and the formation of real property, as is shown by Figure 1 and Figure 2 above. In future there will be a need to separate real property formation from land reform procedures (Figure 3).

The land reform and real property formation require much information of real properties, owners, cadastre, values and land use data from different files, registers and maps. Most of these are produced by order of the National Land Board. Figure 3 shows the stages of information gathering for land restitution and formation:

- production of basic maps
- verification of land use
- verification of legal subjects and restitution decisions
- combination of land reform plans
- real property formation.

Maps, which are used for real property formation are prepared on the basis of existing old maps. The origin is the old cadastre map (1:10,000) from the year 1940. In addition, there are some old land formation maps (1:4,000 or 1:5,000). During the Soviet rule maps were made of kolkhozes, state farms and forest sections. This information has been transferred to aerophotomaps (1:10,000) concerning the Soviet era. The new cadastre map is being drawn on the scale of 1:10,000. These maps cover the whole Estonia in the spring of 1994. There are about 800 map sheets (80cm by 100 cm).

A licensed surveyor produces documents of restitution and real estate formation. The applicant finally gets following documents:

- an extract from the cadastre
- a survey certificate
- decision of the municipal government to return unlawfully expropriated land
- a map of formatted real property.

Registration

In 1940 there were 140,000 farms and 50,000-60,000 parcels in towns entered in the land register. The average size of a farm was 22.4 hectares. The new cadastre shall comprise about 500,000 real properties, including units that are restituted to previous owners. The area of an average unit will be 22.7 hectares including 7 hectares of cultivated land (Vallner *op cit*). According to Tiits (1996), an average sized unit shall consist of 11 hectares of cultivated land and 8 hectares of forest.

At the first stage the National Land Board put the existing data into order and prepared inventory maps of existing real properties. In 1991 they began to maintain a register of municipally used land. During the Soviet rule the aim of land registration was to supply natural-economic data on land for central authorities. The register showed how and by whom cultivated land was used. This information supported the large-scale agricultural production. The change from the Soviet registration system to the new land cadastre is remarkable. The registration of private real properties is very important to the market economy.

The data has been collected in files (registers) and stored in personal computers. Estsurvey has translated the titles of the most important registers into English as follows (Figure 4):

- land applicants and the ground of ownership rights
- house owners in towns and boroughs 1937
- lots (sites) in towns and boroughs
- agricultural collective farms
- countryside land 1940
- title register 1940
- present land users
- soil data
- farms
- classification of administrative units
- land taxation
- cadastre.

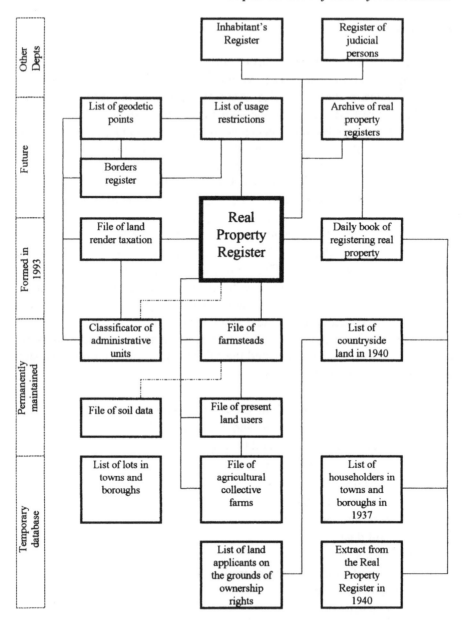

Figure 4   Real property registers (Eesti maakatastri aastaraamat 1992) *(Titles translated by Estsurvey)*

The new land cadastre consists of an alphanumeric land register and a cadastral map. According to the Cadastre Act, the cadastre includes information about real property formation, location and boundaries, the legal status of real property, rights to use real property and relevant restrictions, easements and real property values (Vallner 1993). The cadastral map (1:10,000) is based on old land use maps. This map is drawn manually on plastic sheets. Some pilot projects are testing the use of databases and GIS for cadastral functions.

Problems

About 60% of restitution applications have been handled and decisions about legal owners made, but only about 23,000 real properties are registered in the title register (Tiits *op cit*). The implementation of the land reform is progressing slowly, owing to several reasons. The restitution procedures are complicated. The documents "cruise" between the municipality, the County Land Department and the applicant, which causes delays and unnecessary repetition of work. Because getting the fieldwork started and finding the borders takes most of the working time, much of the work is repeated.

Many applicants hesitated to take their land back quickly because of property taxation and they can delay the process after the decision of the County Commission. It is also possible to cancel the application just before the final decision of the municipality. In such a case all the work has been done in vain. Fortunately the situation has changed according to this.

So far land restitution has been effected mostly in "clear" cases, where it has been possible to give back the land at the same site as in 1940 and there has been no disagreement between heirs. In future, more complicated cases will be common and consume a great deal of time. The replacement of land with land at a new site does not seem to function in practice. No legislation concerning land consolidation exists in Estonia. Obviously land consolidation procedure will have to be worked out after the restitution process.

One of the most important reasons for the slow progress of the restitution process is the lack of resources and techniques especially on the level of local government. There is also an obvious need for legal and administrative staff training.

Some problems are related to the establishment of property markets in areas of high interest (inner cities and surroundings) and to the regulation of urban development and land markets.

During last five years the Land Reform Act has been changed ten times and all related legislation as well (Tiits *op cit*).

It is obvious from what has been said above that the implementation of a land reform in Estonia requires further development of processes, resources and techniques, staff training as well as new technology before the aims can be successfully achieved.

References

Eesti maakatastri aastaraamat, 1992, *Eesti Vabariigi Riiklik Maa-amet*, RE Eesti Maauuringud. Tallinn 1993, 112 pp.

Leväinen, K.I., 1993, *Land Information Management in Estonia.* Unpublished report, Dec 15, Tallinn, 39 pp.

Leväinen, K.I., 1994, Implementation of land reform in Estonia, *Surveying Science in Finland*, **12**(1), 14-26.

Maa-amet, 1993. *Cadastral System in Estonia.* Unpublished memorandum of Estonian National Land Board. Tallinn, 9 pp.

Maareformi täna, 1993. *Maareform täna*, Õigusaktid ja kommentaarid, A/S Kinnisvaraekspert, Tallinn, 122 pp.

Raitanen, P., & Tenkanen, A., 1992. Land reform and cadastre in the former USSR states, *Surveying Science in Finland*, **10**(2), 3-10.

Tiits, T., 1996. Managing director of A/S Kinnisvaraekspert. Discussion 22.11.1996.

Vallner, R., 1993. *Report on Land Reform, Real Property Registration, Geodesy and Mapping Activities.* Unpublished memorandum, Estonian National Land Board, Jun 20. Tallinn, 18 pp.

Vallner, R., 1993a. Interview of Raivo Vallner, Director General of Estonian National Land Board, Nov 10. Tallinn.

Vatsel, U., 1993. Interview of Uno Vatsel, Chief of Department of Land Management, Estonian National Land Board, Oct 19. Tallinn.

Wulff, H., Niederberger, H. & Leväinen, K.I., 1994. *Land Information Management in Estonia.* Phase 1 Report. Tallinn, 148 pp.

GIS and the management of rural land

Julian SWINDELL
School of Rural Economy and Land Management, Royal Agricultural College, Cirencester, England, GL7 6LS

Abstract

The natural landscape can be divided into three broad environments; the unmanaged natural environment, the managed rural environment and the controlled urban environment. The natural environment underlies everything and will always reassert itself if human management ceases. Financially, the urban environment is the most profitable to the professional land manager and therefore accounts for the dominance of urban studies in land management professions.

The rural environment is, however, far more extensive and can be considered to be more fundamental. The urban environment always needs a rural hinterland to support it. This is of crucial importance to the viability of city states, which are dependent upon foreign rural environments for support. It has also been the case in some former communist states, where urban economies have failed, and people are dependent on the supportive rural economy for basic resources.

Because of the financial strength of the urban environment, much development in computer-based land information systems (LIS) has been concentrated in that area. There are, however, strong arguments for developing full geographical information systems (GIS) for rural land management as well.

Units of spatial interest in the rural environment can cover very large areas. The number of people professionally employed in managing these is small, and so collecting information by traditional surveying methods is expensive and slow. Using digital data sources can speed up these processes.

The three environments: natural, rural and urban

The terrestrial landscape can be thought of as three interacting environments, the natural, the rural and the urban. The natural environment can be defined as the world as it would be without the presence or influence of humanity. The rural environment in turn can be considered as the managed natural environment. Those aspects of the natural environment considered desirable, such as soil fertility, are encouraged, and those considered undesirable, such as weeds and pests, are discouraged. The final, urban, environment is a controlled, artificial environment of which all aspects are directed and pre-determined.

The natural environment

It can be argued that there is no natural environment left on the Earth, as all of it is influenced by human activity, even if unintentionally. Ozone depletion in the atmosphere due to CFC aerosols can affect wildlife in any part of the high latitudes. Introduced flora and fauna can fundamentally upset ecosystems that may not yet even have been discovered.

It can equally be argued that catastrophic atmospheric changes and overwhelming disruptions of biogeography due to geological upheavals have been going on for all time, and humankind is just one, fairly new, instrument of change. The definition of a natural environment, in the context of the argument of this paper, is an environment in dynamic, inherent equilibrium with its intrinsic and extrinsic constituents and influences. This means, for example, that the climax vegetation of an abandoned rural landscape is as natural as the pre-settlement vegetation, even though they may in fact be different (Hoskins, 1955).

The rural environment

Any environment which humankind has altered in order to extract resources for their own development, other than in simple hunter-gatherer societies, can be considered rural. The most obvious example of this alteration is the development of agriculture. Originally, useful plants were encouraged to grow in accessible locations and competitive vegetation was discouraged. Irrigation was introduced, healthy cultivars were selected and used for seed propagation and agricultural landscapes developed. In other words, the natural environment came under control, with its development directed in pre-determined ways (Tivy, 1990). Further developments within this environment would include forestry, water storage, mineral and fossil fuel extraction and hydroelectric energy generation. The raw materials and energy for organised society thus come from the natural environment via the managed systems of the rural environment.

The other major resource exploited from the rural environment is land for urban development. Towns and cities are not normally cut out of the wilderness, they arise within rural communities. Exceptions would be mining towns in, for example, gold-rush areas. The commonly short life of such communities emphasises that they are aberrations with no supportive rural context. Towns nearly always arise from an existing, productive rural environment, which can produce the resources the towns require.

The urban environment

People are communal creatures and have always lived within groups of one kind or another. These would originally have been family and kinship groups, then larger communities or tribes and finally political states (Canby, 1962). The late twentieth century witnessed further stages in this agglomeration in the growth of multi-national organisations and multi-state unions, such as the European Union and the Association of South East Asian States. Each growth in the size of group permits greater achievements through increased availability of labour, pooling of talent, more efficient use of resources and increasing social support for specialist artisans, craftsmen and administrators.

For these groups of people to work together it was essential, particularly in pre-motorised societies, for them to live together. This led to the development of hamlets, villages, towns, cities and modern conurbation. Thus the urban environment developed, to provide the context within which the bulk of the population of late twentieth century population lives and works.

The interaction of the three environments

Initially it may appear that these three environments are adjoining. We travel *out* from the city *into* the countryside and further on *into* the wilderness, if we are fortunate enough to live in a region that still has any. In reality, they are all aspects of the same continuum. Part of the natural environment is rural, and part of the rural environment is urban. If human management is withdrawn, the natural environment always reasserts itself.

This can be seen in the fate of the Mayan culture in Central America. For many years it was a problem to archaeologists that this vibrant civilisation appeared to have developed in jungle areas, without significant rural hinterlands (Willey & Sablof, 1974). This was in stark contrast to all other ancient civilisations in the Old World. More recent investigations have shown that the Maya *did* in fact have extensive rural landscapes. However, with the collapse of the Mayan civilisation, the jungles reclaimed the land so thoroughly that the enormous architectural monuments appeared to be little more than hills (Giles & Stuart, 1989) and the agricultural field system had disappeared, almost without trace (Fash, 1991). However, it had been there.

This continuum of all terrestrial environments means that we must consider the effect that land management will have on all of them, whether or not we are trying to think exclusively in terms of urban or rural land management.

The rural environment and its importance

The rural environment is the warehouse from which we draw the essential resources for our lives and cultures. When that warehouse fails, through war, drought, over-exploitation or economic collapse, disaster inevitably follows. The recent famine in Ethiopia brought about by drought and civil war is well known. The consequences of economic collapse can be equally great, if not so readily apparent. A recent conversation between the author and a Hungarian friend illustrated this.

He, the Hungarian, had been to the Ukraine in the early 1990s to visit some distant relatives for the first time. He found that the impact of the collapse of the Communist State had affected people differently. Those living deep in the countryside were having a difficult time, but were surviving. They had land, they could grow food and they got by, probably as well as they ever had this century. City dwellers that had relatives in the country were also managing. They could obtain food from those relatives, but getting supplies was always a problem, and they had little they could offer in return. City dwellers with no rural relatives were suffering the most. They had no food and no means of obtaining any. Their money was largely worthless and once they had traded their material possessions for food there was nothing else they could do. Similar problems can currently be seen in Iraq, where the effects of war and economic sanctions have destroyed the economy and those who have no direct access to the rural environment are approaching near starvation.

The importance of a rural hinterland is also illustrated by the cases of modern City States such as Singapore and Hong Kong. During the Second World War, one contributory factor to the surrender of Singapore by the British forces was the loss of the rural hinterland of mainland Malaya. There would be no water or food for the city and starvation would quickly have conquered it, even if there had been a military possibility to resist longer. Hong Kong was returned to Chinese sovereignty in 1997, even though the actual city is not subject to the lease of 1897, only the New Territories. However, it was felt that without this rural area the viability of the city was impossible and its future therefore lay inevitably with mainland China.

These few examples show the fundamental importance of the rural environment as a resource base for modern communities. Its importance from an amenity and recreational standpoint is also undeniable, both on economic and more fundamental spiritual levels.

Rural land management and GIS

When one examines the current commercial applications of GIS technology to land management, it would seem that urban, rather than rural, managers are being targeted by the software developers. GIS is used for identifying development sites, for managing urban cadastral records, for estimating target populations of consumers for shopping centres, for directing traffic and so on. This is clearly shown in the pilot project for the proposed National Land Information System (NLIS) in Great Britain. All of the pilot areas used are urban and the system is currently aimed exclusively at handling urban property records (Sizer, 1996). This is due partly to the complexities of rural land ownership and tenancy in the UK but also, very largely, to the fact that there is much greater turnover of urban property than rural property, both in terms of the number of transactions and the value of the market. This makes the NLIS potentially much more viable as a 'pay as you use' service in urban areas.

This financial argument for the commercial development of GIS in an urban context is understandable but it does not mean that rural land management could not also benefit, or that it would be an unprofitable market sector. The rural environment is enormous in area, much bigger in nearly all countries than the urban. Its agricultural and natural resources are vital, as argued above. Despite this there are comparatively few professional rural land managers. In the UK urban/commercial land managers out number rural land managers nearly eight times (RICS, 1994). In most European countries there is not even a recognised rural land management profession. Those few professionals who *are* responsible for the management of this enormous resource would benefit greatly from technologies, which would allow them an overview of their area of interest.

Limitations of traditional, map-based rural land management

Traditionally, rural land managers have used paper maps as an important tool in analysis and planning, but they have well known drawbacks. Different thematic maps of the same area are often drawn to differing scales and projections. Large regions have to be dealt with by small-scale maps, or large numbers of unwieldy large scale maps. It is all but impossible to overlay and sieve out more than a few themes at a time (McHarg, 1971).

An excellent example of the limitations of traditional map based approaches to rural planning is the Integrated Administration and Control System (IACS) used in the European Union, by which farmers make subsidy claims under the Common Agricultural Policy (CAP). All farmers making claims have to submit large scale, officially approved topographic maps (Ordnance Survey maps in the case of the UK), hand annotated to show the fields covered by the subsidy claims. In the UK alone there were thousands of individual submissions. Even if one made a conservative estimate of two map sheets needed per claim, there would be tens, if not hundreds of thousands of hand drawn paper maps awaiting inspection. Then they have to be updated at regular intervals. It is clear that any system of control based on such an immense collection of hand written and drawn records is all but impossible. This is exacerbated by the intention of combining these maps (which are not digital) with remotely sensed satellite images (which are digital) in monitoring agricultural fraud (van der Laan, 1994).

If the IACS submissions were handled by digital GIS, their potential changes completely. Field areas can be calculated automatically and updated easily by the inspecting authority itself, and not by the claiming farmer. Satellite and map data can be integrated directly and conflict between claims and reality highlighted immediately. It would also save a huge amount of time for the farmers when preparing their claims, although they would have to invest in having their farm maps digitised.

Remote sensing and rural land management

Brief mention was made above of the use of satellite imagery in agricultural monitoring. This is another instance of the potential use of GIS in managing the rural environment. Remote sensing is much used in urban land management and companies are now selling digitised aerial photography of many cities (Geoinformation International, 1996). This data is undoubtedly useful but has limitations in this environment:

- apart from major building works, which are usually well documented in other resources, little change will be seen in images taken at different times. The ability to repeat survey, possible with remote sensing, is of little benefit;
- the images only show buildings, and not land use or ownership. Large developments may be split between several owners; multi-storey buildings can be in multiple occupancy; a warehouse may be partly converted to other uses, and so on. None of this is easily detectable on the images.

The use of remotely sensed images can be far more productive in the rural environment. Ground cover visibly changes during the course of the seasons, and these changes are significant, which makes repeat imaging important. Land ownership is still invisible, but land divisions are usually very apparent as fences, walls, roads and streams, and these nearly always form the basis of cadastral divisions. Land use and resource potential can often be accurately determined by expert classification. The increasing availability of remotely sensed images makes it easy to keep land use databases up to date by using GIS, even with the comparatively small labour force available for rural land management.

Digital rural spatial data

Two important developments in digital spatial data in the rural environment are the availability of cheap, digital, ortho-rectified aerial photographs and the agricultural use of global positioning systems (GPS).

Geo-referenced ortho-photographs

Satellite images have been employed in large area monitoring for some time, but their low resolution and cost, particularly for contemporary data, make them of limited use for individual farms and rural estates. Paper aerial photographs have been available for many years, but their practical value is limited. The images have to be rectified before they can be used for detailed work and this is beyond the means of most farmers. This has changed with the advent of digitally rectified, scanned photographs. PC based software is available to carry out this rectification and it is comparatively cheap to commission small format photographic surveys. In addition to this, in the UK the National Remote Sensing Centre now markets full colour digital, ortho-rectified and geo-referenced aerial photographs for the whole country through their *Orthoview* service. This means that anyone can obtain digital, geo-referenced images of their own land and work with it on an office PC. Similar schemes are likely to develop in other EU member states.

A common way of using ortho-photos was to overlay them with vector-based topographic and cadastral maps, which were derived from ground surveys. The great potential use of digital ortho-photos for the rural land manager is that they can dispense with costly and time consuming ground survey. Field boundaries, roads, set-aside arable land, even worn footpaths can be identified on the image and vectorised on screen. This means that the vector map can be derived from the image and that they can be designed to show exactly what the land manager needs. Thus boundaries are not undifferentiated black lines but can be identified as hedges, walls or fences. Fields held under different tenancy agreements can be identified as such and linked to databases of farm information.

Agricultural use of GPS

Tremendous progress has been made over the last few years in the discipline of precision farming (Blackmore, 1994). The premise is that agricultural inputs, such as fertilisers and herbicides, can be targeted precisely at those areas of the land, which will benefit from their use. The ability to do this has come about largely through the use of GPS units mounted on agricultural vehicles, which can then be precisely tracked (Auernhammer & Muhr, 1991). The most developed aspect of this technology is the creation of crop yield maps, where crop yield is measured directly by a combine harvester which is being tracked by GPS (Swindell,

1994). This spatially referenced data is then interpolated into a map of crop yield variability with GIS.

An increasing number of farmers are using GPS systems on their vehicles, which means that they can collect spatial data rapidly and easily. This provides an immediate source of management data, which can be integrated with other information, again through the use of GIS and other database management systems.

General purpose GIS software developments

GIS software has a reputation, largely justified, of being expensive and difficult to use. Many educational courses in GIS devote a remarkable amount of their time to just learning how to use the programs, rather than investigating the concepts and theories of spatial data management. The very term GIS has become a problem. Most laymen do not want to touch it because of its perceived difficulties. At the Royal Agricultural College we do not even call our third year option on GIS just that, we call it *Information technology and the management of rural land*, so that we do not frighten students off! (Swindell, 1996)

This situation is changing. GIS functions that are directly useful to farmers and land managers are being incorporated into farm management packages such as *Optimix* and *Farmplan*. More importantly still, GIS functionality is soon to be added to basic business software, such as the *Microsoft Excel* spreadsheet, which will bring it within the reach of all businesses and professions.

Conclusions

The very problems which make the management of rural land difficult; its sheer size; its dynamic, short term variability; the small numbers of professional managers; make it ideally suited to the greater adoption of GIS as an important management tool. The data is increasingly available in usable forms and the software is becoming easier to use and more readily available. Most importantly, GIS, linked with remote sensing can give us a better view and understanding of the whole rural environment on which we are all ultimately dependent on for our succour and survival.

References

Auernhammer, H. & Muhr, T., 1991. The use of GPS in agriculture for yield mapping and tractor implement guidance, *DGPS '91- First International Symposium on Real Time Applications of Global Positioning Systems,* Duesseldorf. **2**, 455-465.

Blackmore, S., 1994. Precision farming: an overview, *Agricultural Engineer,* **49**(3), 86-88.

Canby, C., (Ed.) 1963. *The Epic of Man,* Time-Life International, Amsterdam.

Fash, W., 1991. *Scribes, Warriors and Kings: the City of Copan and the Ancient Maya,* Thames and Hudson, London.

Hoskins, W., 1955. *The Making of the English Landscape,* Hodder and Stoughton, London.

Geoinformation International, 1996. Cities revealed, *Geoinformation International,* Cambridge.

Giles, S., & Stewart, J., 1989. *The Art of Ruins: Adela Breton and the Temples of Mexico,* City of Bristol Museum and Art Gallery, Bristol.

van der Laan, F., 1994. Policing Europe's Common Agricultural Policy, *GIS Europe* **3**(6), 32-35.

McHarg, I., 1971. *Design with Nature,* Doubleday & Co, New York.

RICS, 1994. Chartered Surveyors 1994, Charles Letts and Co, London.

Sizer, P., 1996. The Bristol conveyancing pilot, *NLIS Gazette,* **1**, 3-4.

Swindell, J., 1994. A rich harvest: integrating GPS and GIS on the farm, *Mapping Awareness in the United Kingdom and Ireland,* **9**(1), 32-35.

Swindell, J., 1996. Information technology and the rural surveyor. In: *Proceeding of FIG Commission 2 Joint Workshop: Computer Aided Learning and Achieving Quality in the Education of Surveyors,* 219-228, Helsinki University of Technology, Finland.

Tivy, J., 1990. *Agricultural Ecology,* Longman Scientific and Technical, Harlow.

Willey, G., & Sablof, J., 1974. *A History of American Archaeology,* Thames and Hudson, London.

"Let us build us a city" – the growth of a town in Zambia

Dr. Emmanuel MUTALE

Bartlett School of Architecture, University College, London

Abstract

In its original biblical context, the rallying call "Come, Let us Build us a City" (Genesis 11:4) refers to humanity's self-centred attempt to build a city with its Tower of Babel that reached to the heavens. That this defiant, egotistical and failed attempt to build a city should have been chosen by Kitwe's city fathers for the town's motto either denotes their ignorance of the Bible or a determination to succeed where Babel failed. The City of Kitwe was, from the beginning, a divided town: between the European and African residents, and between the private mining interest and colonial authority. These divisions have affected the physical form and structure of the city through to the present day. Indeed, the very name Nkana-Kitwe reflects its history as 'Twin-townships'.

This paper identifies political, and economic structures, explores their power relationships in influencing the urban development, built form, and urban administration in the City of Kitwe.

The central focus of this paper are the early negotiations over the twin-township structures involving the mining company, the Territorial government, the Colonial Office in London and other territorial local business interests in the establishment of Nkana-Kitwe. After tracing the physical development of the city, and that of its management structures, the concluding section draws together into explanatory theory the interests and power structures in urban planning and management.

Although based upon archival material, the intention in this paper is not to write a history of the City of Kitwe, but to relate that history to the complex of forces, which have affected its urban development and management.

The origin of the twin-townships

The following account of negotiations, derived from the Zambia National Archives (ZNA) and the Public Records Office (PRO) in London, illustrates the uneasy relationships in the development of the twin-townships of Nkana-Kitwe. Although copper was discovered at Nkana in 1905, the mine did not open until 1923 and production only commenced in 1932. By 1929, a mine township had already been established at Nkana, and a management board was constituted in 1935 (Copperbelt Development Plan n.d). The formal basis for 'twin' mine and public township was thus applied retrospectively to Nkana. The African compound consisted of grass-thatched rondavels, European houses of grass-thatched bungalows. For all intents and purposes, Nkana could be classified as a company town defined by Allen (1966) as:

> a community which has been built wholly to support the operations of a single company, in which all homes, buildings, and other real-estate property are owned by that company having been acquired or erected specifically for the benefit of its employees, and in which the company provides most public services.

Anticipating the growth of a non-mining population on the outskirts of Nkana mine, coupled with the mining company's desire to rid itself of the social burden of housing its own employees, the Territorial government decided to co-operate with the mining company to develop a public township adjacent to Nkana mine. While the government preferred a unified town, the mining company, keen to resist local government control, resisted such a proposal and suggested that the company be allowed to develop its own township to meet the requirements of housing and trade (ZNA: RC/1427: dated 27 June 1931). Although the government's proposal for a unified development made economic sense, this rational attempt at planning could not be pushed too hard against the wishes of Rhokana (the mining company), and the Acting Governor could only feebly respond:

> I thoroughly understand the attitude of the mining company in regard to possible control by local authorities, which is the lion in the path of unification of development...We were exploring the possibility of getting some form of unified development without committing the mining company

in any way to render themselves liable to local authorities. We have been only throwing out feelers (ZNA: RC/1427: dated 17 August 1931).

A chief reason for rejecting a unified development was the future possibility of heavy local government rates and taxation against both the mine plant and township, raising production costs and reducing profits.

In 1932, the colonial governor Sir James Maxwell, furnished the colonial Secretary in London with proposed plans for the two townships at Nkana, as a model for all mine areas on the Copperbelt (PRO: CO/795/50/36295: Dated 7 January 1932). Rhokana held land and mineral concession rights at Nkana from the British South African Company (BSA) in five farms of approximately 10 square miles each; three (Farm Numbers 839, 840 and 842) were held in fee-simple, the remaining two (Farm Numbers 841 and 843) on 50 year leaseholds (Figure 1).

Concerning the Mine Township it was proposed that Farm 841 'Kitwe', lying west of the railway and held on a 50 year lease by Rhokana Corporation, should be excised from the parent parcel and conveyed to Rhokana Corporation in fee-simple. This would form part of the mining township and be developed solely for mining and housing for mine employees. The government would not interfere with the development of the township, but would refuse to grant trading licences.

Concerning the Public Township, the following were proposed:

1. the corporation or a subsidiary to undertake to develop and provide services on land east of the railway line (Figure 1) held on a 999-year lease granted by the government, and the former would constitute the local authority under s5 of the Townships Ordinance. The establishment of any other local authority apart from the corporation would only be possible if the local authority agreed to take over the public services at an agreed basis;
2. in addition to making profits from dealing in the land, the governor would consider granting the corporation half the revenue from trade licences and half the fines collected in the public township;
3. the railway company would make its own land arrangements with the corporation, and that no other township would be created by the government within a specified radius of Nkana so long as a considerable portion of this public township remained undeveloped; and

4. lease conditions were to provide the government power to demand the re-conveyance of land necessary for present and anticipated use, with a provision for compensating the corporation on present use value of the land.

The proposals would provide the quickest way of establishing the public township at Nkana, using the financial stability of the corporation and making use (by way of extension) of existing service lines in mine area, and would also relieve the Territorial government of a heavy financial burden.

The Colonial Office treated the matter with grave concern questioning the morality of handing over control of its people to a foreign profit making company for financial reasons:

> it is not our business to increase the profits of a largely American absentee landlord by facilitating their control over our own people (PRO: CO/795/50/36295 dated 9 September 1932).

The Colonial Office was also concerned about potential problems should it be necessary for the government to take over control of the township. In the face of the growing need for a public township at Nkana and yet given the unfavourable financial situation of the Territorial government, the Colonial Office reluctantly agreed to the proposals provided that:

1. the public township to be established by Rhokana on land leased for 99 years preferably 50 years with a provision for compensation in respect of improvements at the end of the lease;
2. Rhokana to appoint the majority on the Township Management Board, the government to be represented by a nominated member.

The Colonial Office did not expect the arrangement to last more than fifty years, considering that it would probably break down even before the expiry of a 50-year lease.

The Territorial government opposed the Colonial Office's modification of the proposals, arguing on behalf of Rhokana that the heavy capital outlay in developing the public township justified a longer lease. Against the feared excesses of company control, it argued that all bye-laws made by the Board would be subject to approval by the governor-in-council, and violation of bye-laws would be policed by a government judicial officer. The Territorial government repeated its inability to incur the expense of developing a public township, and insisted that Rhokana should be allowed generous terms of tenure. This still did not convince the Colonial Office, who directed that Rhokana should be granted a 99 year lease in respect of Subdivision B of Farm number 841 Kitwe (east of railway line), to replace the existing 50 year lease for the development of a public township.

An infuriated Major Pollak on behalf of Rhokana declined the offer, arguing:

> In these circumstances for the government now to offer the corporation the same period of lease as that which any individual can obtain in respect of a residential or business plot is, I submit with all respect, not at all reasonable. If, as I contend, the poorness of the sales effected by government at Ndola and Luanshya was due to restricting the leaseholds to 99 years, what hope would the corporation have at Nkana if it could not even grant a 99 year lease? (PRO: CO/795/62/5587 dated 23 August 1933).

Rhokana now argued that Nkana had since lost its prominence to Ndola (a nearby rival town), so that a public township at Nkana was now no longer necessary, and suggested that trading in the mine area should continue, with Rhokana erecting reasonable structures for rent to the traders. The Colonial Office's refusal to grant Rhokana a 999 year lease diminished the prospect of a public township at Nkana for some time, and the Territorial and Colonial government had to concede to Rhokana's proposal on trade. Still concerned about keeping Rhokana in check, the Colonial Office wanted:

1. Rhokana to apply for the declaration of Nkana as a mine township - a situation which, under the Mine Township Ordinance would prohibit trade in the mine township, except with the permission of the Governor;
2. no trading within the mine area without the approval of the Governor - a condition which could be used should Rhokana not want to apply for the declaration of Nkana as a mine township or before the area is declared a mine township;
3. to limit the number of shops to the existing level, any new shops to seek the permission of the Governor;

4. the Territorial government to undertake to allow trading for 10 to 15 years irrespective of the development of a public township; and,
5. in return for allowing trading activities within the mine area, Rhokana to let houses to government at reasonable rents.

Having settled on the lesser and temporary solution of letting Rhokana build shops within the mine area for rent to the business community, the next sticking points were on the length of the leases to be granted by Rhokana to shop owners and the amount of rent to be charged. Whereas the government favoured a much longer lease, or shorter ones with renewal clauses, Rhokana preferred monthly tenancies or 1 year leases. The government counselled Rhokana that the assurance not to establish a public township might well depend on agreement being reached on the rent and tenurial questions. Rhokana were incensed by what they perceived as a deprivation of use-rights, especially that the land was held in freehold with no conditions attached specifying use. The Colonial Office interpreted Rhokana's objections as aimed at seeking exactly the control, which the government did not wish them to have. Rhokana, who said that they would be justified to eject any trader they found supporting industrial action against the corporation, explicitly accepted this observation. With the death of Major Pollak of Anglo-American Corporation in 1934 (a major player in the negotiations with Rhokana), the government hoped to make progress in its negotiations with Rhokana.

Agreement was finally reached on a number of points including the following:

1. Rhokana to apply for Nkana to be declared a mine township;
2. trading limited to the number of existing shops, any further applications could only be approved by the Governor; and,
3. the government was to allow trade to continue in the mine township for twenty years after the agreement.

Financial constraints could not allow Rhokana to construct trading premises for rent in the mine area, and a suggestion that traders be asked to put the money upfront for the construction of these premises was seized upon by Sir Hubert Young, the new Governor:

If traders were to put up the money, it re-opened the possibility of the traders' stores being erected in a public township and no trading in the mining township (PRO: CO/795/76/25628 dated 27 September 1934).

Following Rhokana's failure to erect reasonable trading premises, the Territorial government approved the continuation of temporary trading arrangements while pursuing a modified form of public township.

Notice 2 of 1935 declared Nkana a mine township under the Mine Township Ordinance. Rhokana maintained that streets in the mine townships were private and reserved the right to close them to the public at any time. After government intervention, the Mine Township Ordinance was amended in 1939 (PRO: CO/795/110/45228 dated 13 June 1939), declaring streets in the residential areas (European section) as public, and the mining plant and streets in the compound (African section) as private, possibly to facilitate the control of African movement in compounds. As a private township, Nkana was exempted from a number of provisions relating to public townships, especially the exemption from general property rates.

In April 1935, the Territorial government in negotiation with Rhokana proposed that the corporation alienates and leases out a limited number of stands, provides the infrastructure, and that the new public township be managed by the Mine Township Board to include an appointed government officer. In return, the corporation was to be reimbursed its expenses on the provision and maintenance of services from the premiums and rentals of the alienated stands. The Territorial government forwarded these proposals to the Colonial Office in August 1935, but the final communiqué differed markedly. For a public township at Nkana (PRO: CO/795/76/45077 dated 17 August 1935):

- stands would be alienated and allocated by the government for 99 years on land acquired from Rhokana Corporation Ltd.;
- stands for trade purposes were to be limited to the existing numbers with a possible additional two stands in each trade sector with preference in allocation being given to existing traders. This restriction would be effective for twenty years from 1 April 1936 and designed to reward existing traders who were to advance the money for the establishment of the public township;

- advance payments for the stands were calculated to allow for the provision of social and physical infrastructure. Subsequent income from the sale of stands and rentals was to be used for the maintenance and extension of the town. Thus, a public township would be built at no expense or profit to the Territorial government. In addition to limiting the number of traders in the public township, the government undertook not to establish another public township within a radius of 16 Kilometres (Ten Miles) of Nkana's smelter stack. This undertaking was for a period of twenty years from 1 April 1936 unless population or other pressures should warrant doing so. In a gesture of goodwill, Rhokana agreed to:
  - give up the land required for the public township without compensation; and,
  - to carry out engineering works and to provide the services in the public township at cost price;
- local residents appointed by the Governor would constitute the Management Board under the chairmanship of the District Officer.

The public township was established under general Notice 397 of 1935 published on 12 September 1935. The layout of the new township would commence in 1936, and trading cease in the mine township on 30 September 1937. The limitation on the number of traders within the public township, and the undertaking not to allow the development of another public township outside it, effectively meant that the public township, like the biblical Jericho, was shut in. Thus, no trader went in and none went out; a condition which earned it the title of a 'closed township' (Kay 1967; City of Kitwe Development Plan 1975-2000; City of Kitwe Street Plan). The township was called Kitwe after the farm on which it was laid out. There is no doubt therefore, that although Kitwe's development was made possible by an initial capital outlay from private traders, the traders, and by extension the initial development of Kitwe were dependent on the mining community. The mining company therefore, was perceived by the Territorial government as a necessary economic ally in the development of a viable public township, with privileges and exemptions from ordinances governing public townships and from charges on land acquired in the public township by the mining company (Notice 397 of 1935). Other possible pointers to the influential position of the mining company are the indemnity granted to

Rhokana in 1934 (PRO: CO/795/68/25528 dated 12 March 1934). This protected the company from being prosecuted for air pollution for 10 miles around the mining area and, more recently, the flexibility with which the local authority has dealt with rates due on mine property. Kitwe public township was officially opened in November 1937 by the Governor, Sir Hubert Young (Bulawayo Chronicle, 26 September 1947).

By 1965, a segregated town had developed in which the two planning authorities (Mines and Municipal council) had each planned and developed separate areas for Europeans and Africans. The African suburbs to the north were separated from the European suburbs in the south by a chain of institutions (hospital, schools, sports ground, transmitter station and police camp) forming a buffer zone. Similarly, other African housing was separated from European housing by a swampy marsh and railway line. Thus, we see racial prejudice translated into physical space. Where African settlement was established within proximity of white areas, it was mainly out of labour needs.

Form and function of urban administration

The primary motive in the establishment of Nkana Mine Township was to provide houses and services for the mine workers free from government interference, and to exercise unfettered control over the work force by tying shelter and services to employment and generally promoting dependence. The township management board was appointed by the mining company and not accountable to the residents in the mine township. Consultation with the people's representatives from the township was merely to sound opinions, check omissions or excesses and generally to lend credibility to an otherwise undemocratic institution, since the management board was not legally obliged to heed such advice.

Africans in the public township of Kitwe (at least until 1956) found themselves under a local authority constituted from Europeans, for whom the town was built and in which the African lived at the former's pleasure. As at 1949, there was no African on the Kitwe Management Board (Clay 1949). By 1956, the only African presence on the Municipal Council was from the African Housing Area Boards and the African Affairs Advisory Committee, each of which appointed two Africans to sit. These were not appointed on the full Council, but rather as observers with no voting powers on the Council's Non-European Affairs Committee - the very Committee which would debate, vote and recommend to the Council, decisions affecting the African in Kitwe (Northern Rhodesia 1957)! Even on the African Affairs Advisory Committee, the African presence was restricted to a token two Africans against seven Europeans. Rather than being equals on the Advisory Committee, the African presence was merely tolerated and was sometimes a source of complaints by some European staff who thought it demeaning that their salaries and conditions of service should be discussed in the presence of Africans (Clay 1949). Only after the change in the electoral law granting equal voting rights in 1963, did Kitwe get its first Zambian Mayor (Mr. J.P.S. Kalyati, 1964-65), and first Zambian Town Clerk (Mr. R.W. Musonda, 1972-76). While the enfranchisement of Africans in 1963 was welcome, it was then that local government became politicised. One of the many examples of political interference in local authority after 1963, is the recent case in which residents, interviewed after 20 houses collapsed following the flooding of the Kafue river, claimed that they had been allocated plots by the Ward Councillor (Zambia Daily Mail, 18 March 1998).

As presently structured, Kitwe City Council has 7 departments each headed by a director thus: administration; engineering services; finance; housing and social services; legal services; public health; and water and engineering services. Each director reports to the Town Clerk who is the chief executive. The council's policy-making body is headed by the Mayor sitting with up to 25 Ward Councillors and 5 Members of Parliament. The council has 7 standing committees namely: establishment; finance and general purposes; housing and social services; licensing; plans, works and development; public health; and water and sewerage services.

Having explored the negotiations, which established the twin-townships of Nkana-Kitwe and the resulting urban form and organisation, the next section draws together the complex political and economic forces that have influenced urban development processes and urban management at Nkana-Kitwe.

An analysis of power structures in the urban development process at Nkana-Kitwe

Pre-independence

Three institutions can be identified in the development of Nkana mine township (Figure 1); Rhokana Mining Corporation, the Territorial Government and the Colonial Office. These three groups, objectives varied with one or coincided with another. For example, it is possible to identify a strong relationship between Rhokana and the Territorial government at certain times but not at others. Such affinities were also influenced by personal relationships in these organisations so that, as personnel changed, so did some of these relationships.

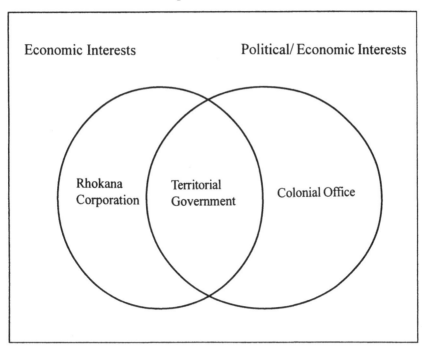

Figure 1 Power bases in the evolution of Nkana
Source: Author

Rhokana's primary objective in the development of Nkana was economic. In order to exploit the rich copper ores in a remote part of the country, the mining company had to provide housing to its workers, recreation and medical care, and other services necessary in a residential area. The combined effect of the nature of mining operations and the company's desire to maximise profits resulted in a highly controlled regime bordering on military discipline. To achieve this objective, Rhokana established alliances with the Territorial Government, which in turn represented its case to the Colonial Office.

The Territorial Government was vested with powers from the Crown to administer the country by the maintenance of law and order, and thus to facilitate the uninterrupted exploitation of resources. The Territorial Government pursued the twin objectives of political control and economic development. Given the dominance of private companies in mineral exploitation, the government's economic objectives were limited to the implementation of fiscal policy, including taxes on companies and individuals. However, on the political front it facilitated the continuous supply of labour and did the bare minimum to provide public services and to protect the native population.

The Colonial Office on behalf of the Crown decided the broad policy framework for the administration of the colonies. Although the legislative council in the territory had powers to legislate, this was subject to the approval of the Secretary of State at the Colonial Office, who therefore, held ultimate veto power.

Although Tipple (1978) identifies the Colonial office in Britain, the Territorial Government, mining companies, missionary societies and South African settlers as wielding political power in Northern Rhodesia, it was mainly the first three institutions which contributed to the growth of Nkana. The trading sector of the settler community, once developed into a critical mass and grown to be financially sound, did also participate in the development of the public township at Nkana. Although missionary activity was present, the religious institutions were silent in affecting the development of Nkana-Kitwe. Their interest in the Copperbelt as a whole included a concern for the changing social and economic values as a result of rapid industrialisation and missionary activity, and the effect these were having on indigenous socio-economic structures (Davis 1933). The failure by the missionaries to influence the growth of the physical environment within which their adherents were expected to live fulfilled lives is regrettable as has also been noted by Davis (1933):

There is an intrinsic lack in a Christianity that widens horizons, creates new visions and stimulates desires, but fails to make its influence effective in assisting men to achieve these ideals.

Similarly, three power bases influenced the growth of the public township of Kitwe: the small time trading community (but big enough to provide capital for the development of the town), the giant mining company and the Territorial Government (Figure 2).

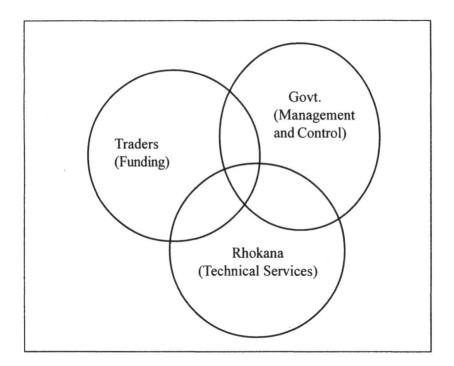

Figure 2 Power bases in the evolution of Kitwe
Source: Author

Absent from this group is the active participation of the Colonial Office notable for its strong presence in the development of Nkana. What was about to take place at Nkana was new and a departure from the norm, and the Colonial Office was wary about what was almost perceived as an abdication of duty to a private company. The small trading community was only providing advance funding in return for a measure of monopolistic gain, and was not interested in the day-to-day management of the town.

We see that the presence of an alternative capital source for the development of Kitwe somewhat softened Rhokana's stand. To remain a good corporate citizen, and possibly to keep open the opportunities for influencing policy related to their mining operations, Rhokana was prepared to surrender their lease on Kitwe farm and put their technical services to use in the development of Kitwe. After the establishment of the public township at Kitwe, day-to-day management was vested in the local management board, which was made up of eminent citizens chosen from among the electorate and public officials - a board that might have included prominent business people from the settler community and public officers.

Post-independence

In the later period following independence in 1964, Kasongo & Tipple (1990) identified three actors in the growth of Kitwe; the mining company, Kitwe City Council and the ruling United National Independence Party (UNIP). To this group must be added squatters whose influence has sometimes been weak and at other times strong, depending on the prevailing political climate. For example, Kasongo & Tipple (1990) cite a case in which Kitwe's settlement plans were substantially changed by squatter action. Absent in this group of actors is the formal trading community, which had played a pivotal role in the establishment of the public township. The activities of UNIP in influencing the growth of Kitwe were mainly restricted to the informal settlements. Tipple (1976) and Kasongo and Tipple (1990) observe that UNIP officials allocated plots in squatter settlements and planned where houses could be built and by whom. Following the introduction of a one-party state in 1972, Mulwanda & Mutale (1994) point to the brutal manner in which UNIP attempted to enforce town planning regulations by demolishing illegal settlements. The re-introduction of plural politics in 1991 saw the squatters wielding more political leverage, which had formerly contributed to the reluctance of the mine authority to force squatters to move from mine owned land to the Twatasha site-and-service scheme between 1974 and 1975 (Tipple, 1978).

66

An example of the informal sector influencing urban morphology in Kitwe is the establishment of Chisokone 'B' market next to the old market and adjoining the Central Business District (CBD), as an illegal structure on land designated for commercial stands. According to Evaristo Onani, the organising Secretary and himself a stall-holder at the same (ZNBC, 1994), Chisokone 'B' was inspired by the Presidential directive not to harass street vendors, who organised a committee to work with the Kitwe City Council, and demarcated the stalls with a plan prepared by the council. The Kitwe Mayor, in a bid to persuade the vendors from occupying Chisokone 'B', told the vendors' committee that the council had plans to build a market complex, but the committee responded by suggesting Chisokone 'B' as the ideal place. To date, there are no indications that the market will move. A task force is in place and this works closely with the civil police to ensure security. Although at the time of this fieldwork (September 1994) the marketeers at Chisokone 'B' had not been licensed, the City Council had asked them to formally apply. The Town Clerk argued that considering the hardships being faced by the people, the City Council could not simply move in to pull down the structures. Daily monetary collections were made by the City Council, regarded as a fine for the breach of council bye-laws. Thus, while formal planning powers still vest in the City Council, political expedience and humanitarian considerations have affected urban development especially in the post-independence period. There is an increasing acceptance of the role of squatters in planning as demonstrated by a recent recommendation for the mining companies to

> negotiate (planning) solutions with squatters, non-governmental support groups, and relevant government departments, especially local government (ZCCM 1997, cited in Hansungule *et al*, 1998).

Conclusions

This historical overview of the function, spatial and organisational aspects of the development of Nkana-Kitwe shows the extent to which colonial institutions, with economic and political interests, contributed to shaping the physical form and the institutional structures in existence today. The mining company and the settler business community with strong economic interests, and the Territorial Government with both political and economic interests, were the initial power bases in the development of Nkana-Kitwe.

The related spatial pattern resulting from this distribution of economic and political power betrays the value judgements of those in power. By the use of cleverly devised instruments, a racially segregated city was developed, the sprawl of which was exacerbated by a broken terrain. Nkana-Kitwe now exhibits a highly compartmentalised form, each compartment contained within some artificial or natural fences, and adding to the cost of services.

The power bases have shifted in influence, with political and executive power having a dominant role in shaping the city's development after independence. Whereas executive power, in the form of the local authority, controlled the growth of the formal sector, political power concentrated its influence in the informal sector. The demarcation of influence has not always been clear-cut, occasional conflicts have occurred, and political attitudes too have changed with changes in the country's political economy (Mulwanda & Mutale, 1994). The independence of the local authority from politics ended with the 1963 local government elections when councils were elected on party tickets. After independence, politicians and council executives became the dominant players in local authority, replacing the economic power of the trading community who had formerly contributed to the development and management of the city. This displacement of economic power by political and council executive power was not complete, but only succeeded in supplanting the small settler economic power. The more formidable mining company continued to exist not only as an influence on Kitwe, but also as a planning authority in Nkana. Nkana Mine Township continues as a distinct residential area within the city under the management of the mining company, and still influences the development of Nkana-Kitwe. With the privatisation of Nkana Division still being negotiated, any discussion about the sort of changes in the administration of the town this privatisation will bring is merely speculative.

This paper has sought to show that after independence, the ability to influence planning decisions is no longer the preserve of formal power structures, but rather includes both the formal and informal structures which have organised themselves around common political or economic interests. Power should not be seen as a monopoly of formal structures, since the informal sector, with no seat on the city's government, managed to influence the physical development of Nkana-Kitwe.

With central government holding 60.3 per cent of the shares in the mining industry, before the on-going privatisation programme (Anon, 1993), the local authority cannot use political leverage to exact economic favours from the mining company, nor does the mining company need to retain a good corporate citizen image. This, and the declining performance of the mining industry since the mid 1970s, has meant that the mining company actually welcomes the shedding of this social burden, although the local authority has been less than keen to take on the management of Nkana Mine Township.

Although the multiplicity of power bases has the potential advantages for checking abuses, and to offer a rich mix or alternative structures of development and organisation, the realisation of these advantages is dependent on the relative balance of power and the degree of co-operation. It also has disadvantages as can be seen in the structure of Nkana-Kitwe where one logical planning unit was split between two authorities thus introducing a break in the planning statement. The actual definition of public and private roads was a source of arguments between the Territorial government and the mining company, with the latter insisting on the privacy of its roads.

The spatial aspect of Nkana-Kitwe shows a relationship between the spatial structure and the distribution of power, initially shared between the mining company and the colonial government. Whereas the mining company's sole objective was economic, the colonial government had the twin objectives (economic and political) of balancing the demands of the mining company and the protectorate role in relation to the local people. The mining company managed to exert influence on the Colonial Office, resulting in the establishment of closed segregated compounds to maintain a high level of control over its workforce. Similarly, the differentiation within functional and residential areas of Kitwe's formal development reflects the ideas and attitudes of the colonial planning profession and strong government control, with residential areas separated on racial lines and clearly defined functional zones. The influence wielded by small private capital was strong enough to win some limited monopoly in trade, and perhaps indirectly influenced the structure of the city through its elected representatives on the management board, but did not directly influence the spatial structure of the city.

In theory, at least, the task of administration was easy, as the mines ensured control of Nkana and the local authority that of Kitwe, but strict control never led to lasting peace. Popular opinion is necessary for a legitimate and lasting administration. An administration by the minority over the many against popular opinion may be forcefully sustained and for a time exhibit a semblance of order, but it will reach a conflict threshold and break-up. An example of this negative policy is the disenfranchisement suffered by the majority African population in local government, either because they were in mine townships or because they lived in non-rateable property (a definition which included all African houses). This exclusion denied the African an early participation in local government, and the chance to influence the course of development in a city which he was eventually going to claim as his own.

Urban management needs to recognise the existence of conflicting forces representing the self-interest of groups or individuals, and to balance these in a way that promotes a creative tension for the benefit of all interest groups. This paper has identified the main interest groups at various stages of Kitwe's development, and how the early tension between the mining company, Territorial Government and small local business community led to the creation of the public township of Kitwe.

In summary, this historical analysis of the evolution and development of Nkana-Kitwe reveals that urban growth and management are both affected by the political and economic interests, and involve a negotiation between various groups. Urban management policies, which fail to identify and negotiate between contradictory interests, are bound to fail. At every stage of Nkana-Kitwe's development, the distribution of effective political and economic power has determined the growth of the city. Where this relationship breaks down, it is because of political compromises made to maintain social stability. Planning, development and management of a city should also recognise the existence of informal structures of power. To treat the development and management of a city as a pure technical exercise diminishes the value, ignores the needs and expectations of the people whom the city is meant to serve, and risks eroding the worth of rational attempts towards a coherent and optimum city structure.

References

Allen, J.B., 1966. *The Company Town in the American West*, University of Oklahoma Press.

Anon, 1993. ZCCM- privatisation's golden opportunity, *Profit*, March 1993.

Bulawayo Chronicle, 1947. *Kitwe has Developed on Novel Lines*. 26 September 1947. Filed at ZNA : SEC1/1528.

City of Kitwe Development Plan, 1975 - 2000, Survey of Existing Conditions, NHA, Lusaka.

City of Kitwe Development Plan, 1975 - 2000, Final Report, NHA, Lusaka.

City of Kitwe Street Plan, Published by Survey Department, Lusaka.

Clay, G. C. R., 1949. African Urban Advisory Councils on the copperbelt of Northern Rhodesia. *Journal of African Administration*, **1**, 33-38.

Copperbelt Development Plan, n.d.

Davis, J. M., (Ed.) 1933. *Modern Industry and the African*, Macmillan and Company Ltd. London.

Hansungule, M., *et.al.*, 1998. *A Report on Land Tenure Insecurity on the Zambian Copperbelt.* Prepared for Oxfam GB in Zambia.

Kasongo, B.A. & Tipple, G.A., 1990. An analysis of policy towards squatters in Kitwe, Zambia. *Third World Planning Review,* **12**(2), 147-65.

Kay, G., 1967. *A Social Geography of Zambia*, University of London Press, London.

Mulwanda, M., & Mutale, E., 1994. Never mind the people, the shanties must go: the politics of urban land in Zambia. Cities, **11**(5), 303-11.

Northern Rhodesia, 1957. *Report of the Committee appointed to examine and recommend ways and means by which Africans resident in Municipal and Township areas should be enabled to take an appropriate part in the administration of those areas*, Government Printer, Lusaka.

PRO: CO/795/50/36295

PRO: CO/795/62/5587

PRO: CO/795/76/25628

PRO: CO/795/76/45077

PRO: CO/795/110/45228

PRO: CO/795/68/25528

Tipple, A.G., 1976. Self-help housing policies in a Zambian mining town, *Urban Studies*, **13**, 167-169.

Tipple, A.G., 1978. *Low-cost housing policies in the copperbelt towns of Northern Rhodesia/Zambia: an historical perspective.* In: Akeroyd & Hill (Eds.), 149-163.

Zambia Daily Mail, 18 March, 1998. *20 Kitwe houses collapse.*

ZCCM, 1997. Nkana mining licence area - ML3, *Environmental Impact Statement, Appendix C, Socio-Economic Issues.* Zambia Consolidated Copper Mines.

ZNA: RC/1427, *The Establishment of a Public Township at Nkana.*

ZNBC, 1994. *Points of View Programme.* Presented by Alexander Miti from Kitwe Studios on 9 September 1994.

A comparative evaluation of land registration and agrarian reform in Austria and Great Britain

Dr. Reinfried MANSBERGER
Institute of Surveying, Remote Sensing and Land Information, BOKU Wien
Robert W. DIXON-GOUGH
Land Reform Research Unit, School of Surveying, University of East London
Dr. Walter SEHER
Institute of Regional Planning and Rural Development, BOKU Wien

Abstract

This paper will consider the systems of land registration and agrarian reform practised in Austria and Great Britain in terms of evaluation, development and functionality. An important factor is the way in which rural land is registered and the relationship between land ownership, land value and agrarian reform. These factors will emphasise the differences between the stable cadastre of Austria, in which the registered boundaries of real estate are legally binding, and the less well-defined yet adequate system of general boundaries used in Great Britain.

Beginning with the Milan and the Franciscan Cadastre developed for the purpose of land taxation, the Austrian Cadastre has been regulated by law since 1969 and, during the last 20 years, the contents of the cadastre and land registry have been converted to a digital format. By comparison, the history of land registry in England and Wales can be traced back to the Doomsday Survey, through the work and status of the Land Registry, to the Doomsday 2000 project, which can be likened to a digital form of cadastre.

The relationship between cadastral systems and agrarian reform will be discussed together with the evolution and management of land use and rural land reform. The historical processes relating to this evolution will be emphasised.

It should be emphasised that practices in Scotland can be significantly different to England and Wales. Therefore, to avoid confusion and where practices differ, those adopted in England and Wales will be described.

Introduction

Nomadic tribes have no real need to measure land and the division of land only becomes a necessity when society has reached a level of settled agrarian development (Richeson, 1966). As societies develop, the systems they adapt for dividing the land, registering those divisions and implementing the resulting agrarian reform emerge. As a result of these factors, cadastre and agrarian reform have different historical roots and a different evolution in the both Great Britain and in Austria; the systems evolving to satisfy the national requirements of regulating the ownership of land and producing sufficient quantities of food to feed the population.

Due to this process of evolution, in respect to both the different legal systems and agricultural structures of the two countries, the realisation of cadastre and agrarian reform has taken two distinctly different pathways. Within the paper, the milestones of historical development and the main differences of land registration and agrarian reform will be identified and discussed.

At a time when international co-operation on agrarian reform is being actively encouraged, particularly throughout Europe, comparisons need to be made between systems, not only to identify the major differences but also to identify the common elements. It is from these common elements that active co-operation and understanding may commence. In 1973, Great Britain became a member of the European Union and Austria joined in 1995. As fellow members, both countries now have to observe a common set of European Laws and regulations, particularly within the special field of agriculture, as well as their national laws and regulations.

One of the most significant aspects relating to recent European agrarian reform is the growing awareness of the importance of ecological considerations of agricultural food production. This is mirrored by a decline in the commercial and economic considerations, which had been of paramount importance since the early 1940s. The gradual implementation of this new 'thinking' (ecology instead of economy) could be achieved by parcel-based subsidies and therefore this factor will have an influence upon both land registration and agrarian reform. In order to achieve the required agrarian reforms, the systems adopted by both countries need to be adapted. A comparison of the systems employed in both countries can identify the relative merits of the two and can be a useful base for an improvement of the respective systems.

With respect to the registration of land, both countries are moving away from systems of analogue land registers and analogue cadastre to digital databases. In addition, there is an increasing awareness that other parcel-based information should be converted to a digital format. This leads to the final step in the development of land registration and agrarian reform, the development of a Land Information System (LIS) to enable all parcel-based data to be fully integrated and available for queries.

Simplistically, the Austrian system may be defined as a fixed or rectilinear system whereas that adopted throughout the UK may be defined as a general boundary system. Since many cadastral surveyors have little experience of general boundary systems, it is worth a few words of explanation concerning their characteristics. Dale (1976) defines a general boundary as

> a boundary whose precise line has not been determined.

The general boundary reflects the situation as it exists and it is normally the responsibility of the purchaser of the land to ensure that the boundary is located approximately where it purports to be on the legal deeds of the property. In order to gain an understanding of the differences between the two systems, it is necessary to consider their evolution.

Definitions

This section contains, in a very general form, the definitions of the most important terms used in the paper. Due to the different systems employed in Austria and the UK, a detailed description is not sensible at this stage. This will be made in the following chapter concerning technical, organisational and legal aspects of the systems.

Land Register. A Land Register is normally an up-to-date and ownership-based register, containing all the information about the interests of lands and buildings (ownership, rights, restrictions, responsibilities, mortgages etc.).

Cadastre. A Cadastre is normally a parcel-based and up-to-date information system, concerning the information about the parcel (geometric description of the extent of the parcel, location, co-ordinates of boundary points, area, taxation value etc.).

Agrarian Reform. The term Agrarian Reform includes all regulations that lead to a new structure of rural land. Within this paper only the measures of the agrarian reform will be considered that have a relationship to cadastre and land register.

Land Valuation. Valuation of land related for taxation purposes. The estimation is parcel-based.

Historical aspects

One of the major differences between the history of cadastre, land registration and agrarian reform in the two countries lies in the way in which it has evolved. In England and Wales, the lack of a constitution has led to a more gradual evolution of the processes than in Austria, which has a clearly defined constitution. The evolution in England and Wales is further confused by the multiplicity of Acts, Commissions and events related to land registration and agrarian reform. This is evident, particularly in the nineteenth century, in various Commissions set up to consider Land Registration in Great Britain. The same is true with agrarian reform, this being implicit rather than stated in most Acts of Parliament related to the countryside. Similarly, the development of the two countries can be contrasted. Great Britain was a very early example of an industrialised country supported by a large population. In contrast, Austria with a much smaller population developed its industrial base at a much later date.

For the purpose of this paper, the origin of land registration and agrarian reform in England and Wales can be dated back to the Norman Conquest of England in 1066. As a result, the Domesday Survey was conducted in 1086, to provide documentary evidence of a fiscal survey, essentially recording the dues owed to the King. Since the Domesday Book is still in existence, it provides a fascinating insight into the land and living conditions of that period.

The evolutionary process of land reform commenced in England in 1215 with the signing, by King John, of the Magna Carta. In forcing the King to sign this document, the Barons were not strong enough to oppose him but had to seek the aid of the church and all classes who had been oppressed by the Crown. The terms of the Magna Carta were of limited interest to 'the people of the nation' in the thirteenth century, but owing to the economic and legal evolution of the next three hundred years it came to embrace the descendant of every villein in the land, when all Englishmen became in the eye of the law 'freemen' (Trevelyan, 1988).

This evolutionary process in Great Britain continued with the Black Death of the fourteenth century and the definition of the Rights and Prerogatives of the common man in the 15th century, giving rise to the voice and needs of the rising middle classes. The War of the Roses eventually led to the breakdown of feudalism, which was essentially due to the decline in the number of villain tenants. This, in turn, resulting in competition for workers amongst manorial lords, bringing into being the class of agricultural labourer prepared to hire out their service to the highest bidders. A further evolution came during the middle of the fifteenth century when payments were made in form of ground rents for their holdings rather than in the feudal burden of personal services.

As the social status of villains improved, their tenure became known as Copyhold. This expression was the result of their title to land being recorded in Court Rolls, the villain tenant being said to hold 'by copy of the court roll', the copy being his title deed. Since all transactions were recorded upon the role, it provided conclusive evidence of the copyholder's rights (Rowton Simpson 1984). In the nineteenth century, Torrens suggested that the British missed the opportunity of setting up a National Land Register based upon the Copyhold Laws. By that time, several statutes had been passed to encourage the voluntary extinguishment of the copyhold tenure. Copyhold was compulsorily abolished in 1925, tenures being transferred as a result of the Land and Property Act.

During the sixteenth century the population of England and Wales grew faster than in any other European country. It has been estimated that the population of England rose from 2.5 million in the year 1500 to about 4.1 million by 1600 (Koenigsberger *et al* 1989). Whilst these estimates may not be accurate they do, however, give an indication of the trend of population rise not only in Great Britain but across the whole of Europe. In practice, such a rise in population meant that with increasing demands for bread and meat, for wool and flax, and for building materials there had to be improvements in farming methods. These came in the form of crop rotation and the extension of agriculture into less fertile or accessible lands.

In England, wool prices increased with demand more rapidly than grain prices. Common land was enclosed and villages depopulated to make room for sheep. A Royal Commission was set up by Cardinal Wolsey to inquire about enclosures. These steps made the problem of food even more severe. During 1550, the English coinage was re-valued causing a severe drop in the export of wool and it became more profitable to enclose land for arable cultivation rather than for sheep farming.

An increase in wealth of Great Britain, brought about by increased industrial production and its social structure, lead to the 'improving landlord' who had the capital to experiment with agrarian matters. Thus, interest in agrarian reform during this period was far more widespread in England than in any other European country. This process led initially to the establishment of numerous small farms, but these were eventually absorbed into larger holdings (Butlin, 1982). This period led to the emergence of a land-owing class who had the leisure, the capital and the interest to improve their estates. This, in turn, paved the way to agrarian reform, which was dependent upon the work of the surveyor. Improvements ranged from increases in market gardening (particularly in Kent and Essex) to serve the needs of London, land reclamation, the increased use of root crops, land drainage (such as the schemes of the Duke of Bedford in the Fens - The New Bedford River), and the planning of parklands around country estates (Maland 1983).

The system of rural land ownership and land use in Austria - the domain system - dates back to the Franco-Bavarian colonisation between the 6th and the 12th century. Agricultural land was owned by landlords, the majority of peasants were land tenants with the duty to pay tithe and to work in domain land. The prevailing form of cultivation was triple-field-husbandry - yearly rotation of winter crop, summer crop and fallow land on the same parcel - pastures were used in common by the village community. In many domain forests peasants were provided with wood and grazing rights. Especially in the alpine regions they produced mainly for their own consumption with little market relations.

This system proved to be more or less stable until the middle of the 18th century when the absolutistic government began to exercise an influence on economic development. Due to fiscal reasons and an increasing demand for goods to supply the army and the court manufactures were founded leading subsequently to an increasing demand for workforce. A reform of administration implemented in 1756 reduced influence and power of the landlords. In 1781 Emperor Josef II abolished the peasant's personal subservience to the landlord and guaranteed freedom of movement and free choice of profession. In the same period first attempts of agrarian reform were undertaken. The imperial act concerning the division of common land in 1768 and the encouragement for an improvement of cultivation methods carried out by 'agricultural associations' established for this purpose, however, turned out to be less successful.

In urban areas the registration of ownership increased at the beginning of this millennium. The registration into so-called 'Stadtbücher' and the issue of deeds gave the citizens security of their real estates. In Vienna the finance laws of Rudolf IV (the donor) became the legal basis for the land register system in the year 1360 (Kandutsch 1995).

In Austria during the eighteenth century (1718 to 1769), the government registered, for the first time, all buildings and all yielding land for financial purposes within the Italian provinces, which then belonged to Austria. The maps for the so-called 'Censimente Milanese' (Milan Cadastre) were produced. In addition, the land was evaluated with respect to three classifications (good, medium and poor soil) and the taxation of the land was subsequently based on its net yield. In 1817, Emperor Francis I of Austria, ordered through the 'Grundsteuerpatent' (Land Taxation Law) the surveying of the whole monarchy. Between 1817 and 1861, approximately 50 million parcels in 30,000 communes were surveyed and mapped. The surveying of approximately 300,000 km² was achieved using the plane-table method. The maps of the 'Franciscan Cadastre' were based on triangulated points and with this the parcels now got a 'stable' geo-referencing. For valuation purposes the land within the 'Stabile Cadastre', as the Franciscan Cadastre was also named, was classified into 14 land-use categories (e.g. field, garden, meadow, vineyard, forest, lake). Each category, specifically for each commune, was subdivided into different classes of net yield.

During this period in Great Britain, the industrial revolution was taking place. Prior to the industrial revolution, the principal status of power was land and, in order to keep land, many of the families of the large estates tied up the land through inheritance to prevent it from being sold by future generations. The key value of the estates depended upon them being kept intact and their total value represented a status that was greater than the sum of their parts.

Event in GB	Year	Event in Austria	*Other Events*
Domesday Survey 1086 The Magna Carta 1215	1050		Norman Conquest 1066
Enclosure Acts	1600		Population changes GB
	1750	Establishment of the Censimente Milanese (1718-1769)	
			Industrial Revolution GB
	1800	Grundsteuerpatent 1817	
Royal Commission 1830 General Register Bill 1830 Mapping of GB by OS (1853- 1893) Royal Commission 1847 Royal Commission 1857 Land Registry 1862 Land Transfer Act 1875 Compulsory Land Registration Act 1897	1850	Establishment of the Franciscan (Stabile) Cadastre (1817-1861) Land Register Law 1871 Land Reform Laws 1883 Evidenzerhaltungsgesetz 1883	Peasants become free owners of land in Aus. 1848 Contract Austria – Hungary 1867
Finance Act 1911	1900		Introduction of Death Duties in GB (1911) First World War 1914-1918
Land Law Legislation 1925		New Constitution (1929)	Wall Street Crash Second World War (1939-1945)
Town and Country Planning Act 1948	1950		Increased need for food production and self-sufficiency
Entry to EU		Surveying Law (1969) Soil Estimation Law (1970) Establishment of GDB Entry to EU	
	2000		

Figure 1 Land register, cadastre, agrarian reform and land valuation: historical milestones

Ownership represented influence, which governed many aspects of local life. The industrial revolution changed this and shifted power to the factory owners who required land both for factories, their workers houses and for their own status.

The sale of land was complicated because of the English Land Laws, time-consuming and resulted in an expensive process of private conveyance and the risk of fraud. These complications lead to the decline in land values at a time when the market for land was increasing. Concern at the increasing confusion and insecurity of title led, in 1828, to the appointment by Parliament of a Royal Commission to inquire into the law of England concerning Real Property.

In Austria the liberal revolution during 1848 resulted in the abolition of the domain system and the liberation of land from feudal service. Peasants turned from being tenants to free owners of their land. However, they had to pay a third of the cost of the land to the former owner. The resulting financial needs forced the peasants to increase productivity of their land. The development of new cultivation methods, in contrast to the out-of-date system of triple-field-husbandry and further fragmentation of agricultural land, caused by a law in 1868, which implemented unrestricted inheritance as well as free division and sale of farms and farming land, gave strong reasons for the necessity of agrarian reform measures. On the other hand the consequences of the inheritance law and sinking grain prices, as a result of massive grain imports from Russia and the United States, led to a great drain of agricultural workforce forming the human base of the industrial revolution in Austria.

This period also witnessed several significant steps in the registration of land and the consequent agrarian reform in England and Wales. The 1830 Royal Commission resulted in four reports, the first two devoted to land registration (1829 and 1830), the third to land tenure (1831-1832), and the fourth to wills (1832). In their main report, the Commissioners concentrated upon the gross defects of the prevailing methods of land transfer, and on the need to establish 'a General Registry of Deeds and Instruments relating to land'. They recommended that a Deeds registry should be established with a general office in London serving England and Wales, and regional District Registries. The Commission took it for granted that the registration of land would be compulsory and this was accepted at the time by Parliament! Had this been achieved, the register of land in the United Kingdom might now have been complete (Riddall 1983). This resulted in the introduction of a General Register Bill (1830) in the House of Commons, which was opposed, largely by solicitors who feared a reduction in their business. Despite the failure of the bill, it was a pioneering document in that it made reference to the Registration of Title as distinct from the Registration of Deeds.

Apathy on the part of the public and opposition on the part of the legal profession had resulted in no progress being made on the registration of land. This led to the report of the Registration and Conveyancing Commission, published in 1850. In 1846, a select committee of the House of Lords came to the conclusion that the marketable value of real property was being seriously diminished by the problems associated with its transfer. A Royal Commission was appointed in 1847 to explore the most effective system for the registration of deeds to simplify and reduce the costs of conveyancing. The Commission concluded that the existing rights to land were too complex for the title to be guaranteed to a public officer but recommended, however, the setting up of a National Deeds Registry. The recommendations of this report form the basis for current land registration in England and Wales. The previously requested deeds register was rejected in favour of the more radical concept of a General Registration of Title to Land.

In 1853 the Ordnance Survey commenced the 25 inches to the mile survey (approximately 1:2,500) of rural areas and 50 inches to the mile of urban areas (approximately 1:1,250) in England, Wales and Scotland. This was the first systematic mapping of 'visible' boundaries in Great Britain and was not completed until 1893. It did, and successive series still do, form the basic tool for Land Registration and agrarian reform in Great Britain.

In 1862, the Land Registries of Scotland, and England and Wales came into being. Registration was to be voluntary and the records to be kept secret. Only the owners of estates and interests were to be allowed to inspect the register and then only their own property. Further recommendations lead to the 1875 Land Transfer Act which was voluntary and once more, not a success. In an attempt to make land registration more attractive to land owners, three new elements were introduced; land had to be sold before being registered, boundaries no longer had to be accurately surveyed but recorded as 'general boundaries' and, thirdly, partial and equitable interests could no longer be registered. The definition of a general boundary means that the exact line of the boundary is left undetermined. Instead the system makes clear where the parcel is situated in relation to certain clear and visible features, such as hedges. This was a reintroduction of a rule that had been applied to conveyance for centuries. The Land Registry was not empowered to make its own maps, the Land Transfer Act simply defining that the applicant provided an extract from a local map that had been defined as the 'public map of the district'. It was, therefore, reliant upon Ordnance Survey maps. In the absence of these, tithe maps and estate plans were used.

The year 1871 was the beginning of a countrywide land register in Austria. In this year, a law was enacted which regulated the establishment of the Austrian land register. During 1883, three principle laws were enacted concerning land consolidation; the removal of agricultural enclaves in forest land; the division of agricultural communities (common land), and adjustment of common land use rights. Together with a Melioration Law, passed in 1884, they paved the way for agrarian reform as a governmental policy. The implementation was carried out by judges and administration officers, so called 'local commissioners of agrarian reform', assisted by a technical department. In 1891, the first land consolidation was completed in Lower Austria and from that time on it had a steady growth until the beginning of the First World War.

1883 was also an important year for the Austrian cadastre: the 'Evidenzhaltungsgesetz' was enacted. This law ordered that all changes of parcels (boundary and ownership) must be entered on the cadastre maps so that the maps are kept constantly up-to-date. Since that time, there has been a constant process of revision of the Austrian cadastre.

The great estates in Great Britain peaked during latter part of the nineteenth century, when a survey of land tenureship (often referred to as the New Domesday Survey) was made during 1873 (Dwyer & Hodge, 1996). This survey showed that in England and Wales, estates larger than 1,000 acres occupied more than 50% of the land area and were owned by some 4,200 owners. In addition, those estates of over 300 acres covered more than 66% of the remaining land area and were owned by some 14,000 owners (Orwin & Whetham, 1964). Ownership of the land was even more concentrated in Scotland and it was claimed that 25% of the land area was owned by some 12 landowners. Thus during this period, the majority of the land was farmed by tenants and less than 10% of available land was either farmed by owner-occupiers (Northfield, 1979).

The Compulsory Land Registration Act of 1897 was introduced for selected areas, initially in the County of London and extended to the City of London in 1902. In order to pass the Act through Parliament, two important amendments were necessary. Essentially, the Act called for the compulsory registration of Title on the sale of a property, or upon granting a lease for forty years or more. The first amendment allowed the Act to be introduced as an 'experiment' into the Administrative County of London. The second amendment came with an Order making registration compulsory only in areas where the County Council requested it. The Register was to remain secret. The 1897 Land Transfer Act made the Ordnance Survey map become the basis of all descriptions of registered land. The map became dominant, so that if any verbal descriptions of the property disagreed with the plan, it was the graphical description that was to take legal precedence.

Until 1898, Ordnance Survey Officers had run the Survey and Map Department of the Land Registry. This responsibility was now passed to the Chief Registrar. Thus, any building developments in compulsory areas would usually be followed by conveyance. This would entail the revision of the Ordnance Survey plan. This created an argument between the Ordnance Survey and the Land Registry over whose surveyors were to carry out the revision work.

In 1910, the Finance Act for the redistribution of wealth from primarily landed gentry was put before Parliament. It provided a valuation on all land in Great Britain and this is still maintained by the Inland Revenue but kept as confidential records. This was referred to as The People's Budget introducing amongst other things, 'death duties', later to be superseded by the Inheritance Tax. This Act, which included the introduction of income tax, led to a general decline in the wealth, status and of the influence of the landed gentry. This gradually led to a decline in the financial position of the great estates and in a deterioration of their maintenance. However, the transfer of property from the ownership of those estates did not reach any significant level until the demand and value was revived during the period between 1911 and 1914. During and immediately after the First World War, many large estates had to be sold and broken up to pay for death duties brought about by the decease of two or three generations in rapid succession. It has been estimated (Thompson, 1963) that between 1918 and 1921, between six and eight million acres changed hands in England and Wales, some 25% of the total land area. This lead to a breakdown in systematic agrarian planning for virtually the remainder of the 20th century. Thompson *op cit* commented that these land transfers

> marked a social revolution in the countryside, nothing less than the dissolution of a large part of the great estate system and the formation of a new bread of yeomen.

The 1925 Land Law Legislation was the result of recommendations made by the Acquisition and Valuation of Land Committee, which met in 1919. The principle recommendation of the Committee was that the provisions of the Act of 1897, under which compulsory registration was impossible except at the instance of a County Council, be repealed on the grounds that the gradual extension of compulsory registration was of a national rather local interest. The 1925 Law and Property Act lead to the gradual introduction of compulsory land registration across England and Wales. The spread was, to a large extent, hindered by the revision of the Ordnance Survey's large-scale plans, which were essential tools for land registration purposes. Despite subsequent legislation in 1960, a considerable number of land parcels remained to be registered.

The Wall Street Crash was the final straw for many major landowners in Great Britain. The following Great Depression led to a steep decline in the value of agricultural land and the first major economic impact on an international scale. Land values only recovered during the latter part of the 1930s when it was necessary for the nation to feed itself and when subsidies were offered to farmers to put marginal land into production and to increase the yield of the land.

After the breakdown of the Austro-Hungarian Empire, Austria had lost the main agricultural regions in Hungary and had to increase productivity of its own agriculture. To provide for self-sufficiency in food supplies, the republican government pursued a protective farming policy by fostering re-settlement of small farms and agricultural co-operatives. Protective duties for grain and grain products were adopted and fixed prices for agricultural products were guaranteed to the farmers. Agrarian reform measures were carried out, not only in the lowlands but also extended to alpine areas using the legal and administrative base left over by the monarchy. Since 1926, an arrangement between the ministry of agriculture and the ministry of economy and justice ensured the mutual free of charge data-transfer between agrarian reform authorities and cadastre or land register.

In 1929, Austria was given a new constitution, which included the measures of agrarian reform. In 1932, two federal laws were enacted concerning land consolidation and land improvement by rural infrastructure. Though they have been modified, these laws still today form the legal basis for the most imporant measures of agrarian reform. In spite of massive support, the economic situation of many farmers got worse during the 1930s, which led to an increasing attraction to the emerging national-socialism. From 1938 to 1945, agrarian reform was performed according to German laws and agrarian reform authorities were incorporated into the German administration system.

The Second World War left Austria's agriculture devastated. The quick recovery of the agricultural sector after 1950 was made possible by the international support of the Marshall Plan. Increasing mechanisation of cultivation raised the importance of agrarian reform measures. During 1950, the procedures of agrarian reform measures as well as the organisation of agrarian reform authorities were put on a new legal base. On the one hand there was land consolidation in the intensively cultivated areas of eastern Austria and in the alpine valley regions. On the other hand, in mountainous areas the construction of rural roads to connect mountain farms to major roads situated in the valleys proved to be the most important and effective measures of agrarian reform. Agricultural statistics show a steady increase of land consolidation measures until 1980. Since then agricultural surplus production and increased environmental sensibility has led to a reduction, not only in quantity of output, but to a qualitative improvement by including ecological planning tasks.

A countrywide estimation of soils in Austria was commenced during 1947 during which the quality of soil is derived by parameters of soil condition, relief, climatic conditions and water conditions. Finally the yield value of the land will be determined by comparing the quality of soils with representative examples of soils within the commune.

Since 1969, Austria has had a Surveying Law. These regulations established a constitutional foundation for the functions of surveying authorities and extended the parcel cadastre to a boundary cadastre, which represents a legally binding proof of boundaries of parcels.

During the last two decades, both the cadastre and land register in Austria has been converted into a digital real estate database (Grund-stücksdatenbank). In addition to all the benefits of digital databases (access, queries, etc.), the redundant information within the cadastre and land register (cadastre being a complementary register to the ownership register and vice versa) has been eliminated.

The historical evolution of land registration, cadastre and agrarian reform shows different roots in both countries and the methods are strongly influenced by external influences. Throughout the last century, the history of both countries and also the requirements of cadastre, land registration and agrarian reform have been similar. This is also relatively true in the ways in which these elements have been realised.

Technical and organisational aspects

To compare the methods and organisation of Land Registration, Cadastre, Agrarian Reform and Land Valuation in both countries would be an extensive venture that would involve a multiplicity of professions including agronomy, lawyers, soil experts, etc. It is, therefore, the intention of the authors to concentrate on those aspects that concern directly to the surveying profession.

Land registration and cadastre

There are common elements of land registration and cadastre in Austria and Great Britain. Both countries have a Land Register, which may now be examined by all, and both countries have a means of linking the mapped parcel (cadastre) to the legal title of the land. This provides an authoritative public record of land ownership and rights. In Great Britain, the term cadastre is rarely used since the geometry of the parcel is not fixed geometrically but defined in more general, graphical terms being based upon maps of the Ordnance Survey. The scales of these maps are 1:1250 and 1:2500. The relationship between the Ordnance Survey and the Land Registry is a complicated one - the sub-division of 'new' parcels being made either by non-licensed surveyors or members of the legal profession. The surveyors of the Ordnance Survey are then delegated by the Land Registry to 'map' the region and this provides the general boundary details. Although Ordnance Survey plans form the basis for the general boundary map (cadastral map), no ownership or boundary feature is implied. It is worth noting that the land register is still incomplete and is likely to be so for many years.

In Austria, the cadastral maps are produced and maintained by the Austrian Federal Office of Metrology and Surveying. The map scales are very similar to those of the Ordnance Survey (in general 1:1000/1:1440 for urban regions and 1:2000/1:2880 for rural regions). In both countries surveyors record boundaries and boundary changes, the major differences between the two countries being that there are no licensed surveyors in Great Britain. The ownership of parcels are recorded by the Land Registry Offices.

Both countries are currently reorganising their Land Register and Cadastre into a digital format. The Ordnance Survey now has a complete digital coverage of Great Britain (LandLine data) and this will eventually assist in the digitisation of all Land Registry information. At present, an individual number identifies data concerning a particular parcel held by the Land Registry. The property register identifies the geographic location and extent of the property. It will also include a plan of the property (www.open.gov.uk). Although all titles held by the Land Registry were expected to be digitised by 1998, it is not possible to make an automatic link to the Ordnance Survey's LandLine data. As a means of moving towards a system of simpler access to land and property information, a feasibility study is currently being undertaken in the development of a National Land Information System. This was predated by the Domesday 2000 project managed by the University of East London. The project provided a demonstration system to illustrate the potential of integrating land and property information. The NLIS Steering Group (NLIS SG) is responsible for overseeing the feasibility project and field trials are currently taking place in Bristol concerning the combination of the records of the Land Registry, the Valuations Office and the Ordnance Survey through postal addresses (www.ordsvy.gov.uk).

Austria is ahead of Great Britain in this respect and has completed the conversion of the Ownership Register and Parcel Register to a digital format. The integration of land register data and cadastre data in the 'Data Base of Real Estates - Grundstücksdatenbank (GDB)' was finished at the beginning of this decade. The overall term 'Real Estate Data Base' means, in fact, an aggregation of different databases which hold information on subjects, rights and objects of real estates (Austrian Federal Office of Metrology and Surveying 1997). The graphical representation of parcels, the Digital Cadastral Map (DKM), should be finished within the next five years. Currently about 65% of Austrian cadastral maps are already digitised. Simultaneously to the conversion from the analogue maps to a digital representation of parcels the boundaries are revised by photogrammetry and by reference to partition maps. In the future the DKM will be a very suitable base for a country-wide Land Information System in much the same way as the Ordnance Survey LandLine data will be to the NLIS.

In identifying the differences between the systems adopted by Austria and Great Britain, respectively, the regulation by law of land registration and cadastre in Austria must initially be considered. Land registration and cadastre are organised on a regional basis in districts. Both registers are public and the data contained within the registers are legally binding (boundaries of parcels in the cadastre, ownership of parcels in the land register). People have a duty of notifying changes of ownership. A public authority will register the modification. If the parcel is divided, the partition must be based on a plan of a surveying authority. Because of these conditions, the land register is guaranteed by the state and people can trust the data contained in the register. The surveying authorities resolve boundary disputes by re-establishing the boundary using the co-ordinates recorded when the boundary was originally established. This statement applies only to parcels stored in the 'Boundary Cadastre', which commenced in 1969. In all other cases, the authority for resolving boundary disputes is the High Court, a similar situation to that encountered in England and Wales, where all matters related to boundary disputes will be heard in the High Court. In these hearings, the surveyor will act as an 'Expert Witness' to assist the court and to be a witness of opinion and fact, together with members of the legal profession.

Additional information such as the area of the parcel, the land use of parcel parts, the information of ownership of land contained within the analogue cadastre (before the establishment of the GDB) etc., are also implied in the Austrian cadastre but are not legally binding facts. As mentioned above, the land register and cadastre are public. Since the implementation of the digital registers, the procedure of obtaining information has changed. Everyone in Austria can access data, subject to charges, using modern communication technologies. Alternatively inquiries can be made at the regional authorities (the cadastral or land register office). This situation is similar in Great Britain, where the Land Register has been open to the public since 1990. For a charge the Register may be inspected, or a request for a search made by post, and a copy of any registered title obtained through the relevant District Land Registry.

Agrarian reform

Whilst the historical evolution of agrarian reform has been quite different for the two countries, the general conditions of agrarian reform throughout the twentieth century have been very similar. During this century, the priority of all measures within this field of activity lay in the increased need for production and self-sufficiency. During the last twenty-five years this has largely changed, and this has been brought about by the following circumstances:

- both countries are now members of the European Union, the United Kingdom of Great Britain and Northern Ireland joining in 1973 and Austria in 1995. As a result of overproduction, and as means of reducing farming subsidies, farmers are not longer dependent purely on the quantity of production. The importance of the environment and ecology has increased in all considerations of European farming policies. The situation is that agriculture is now being encouraged to re-adjust the production of food in the sense of sustainability, preservation and conservation of the environment;

- secondly, there are changes in the economy of both countries. If the statistics of the respective Gross National Products (GNPs) are considered, it may be seen that the agricultural influence upon the economy of European States has decreased in a dramatic manner. The economies, particularly of Austria and Great Britain have changed from a rural to an urban emphasis and with it, the political influence has shifted from the countryside to the town;

- finally, the farming industry has witnessed an increased use of technology and the application of regulations within the last three decades. Both factors have led to a decline in agricultural workforce brought on partly through increased mechanisation but also, in the case of many farmers in Great Britain, as the result of Health and Safety legislation imposed by Europe. This latter factor, has led to the increased reliance upon a specialist, self-employed work force contracted for specific tasks, particularly in the case of the larger farm units. In contrast, however, conservation practises encouraged by European subsidies has seen the emergence of a skilled rural workforce capable of traditional skills.

The only difference in agrarian reform between the two countries, lies in strict regulation of measures in Austria. Land Reform is laid down in the Constitution (§12) and the regulations for the realisation of agrarian reform are defined in Provincial Laws. The responsibility for the implementation of agrarian reform measures lies with the Agrarian Reform Authorities, who are occupied with planning and matters of law. They consist of an Engineering Department (rural and agrarian engineers, land surveyors and landscape architects) and a Legal Department. The measures are planned, financially supported and controlled by the Federal Authorities and are initialised at the request of farmers. Within the numerous measures of agrarian reform, Land Consolidation and Consolidation by Voluntary Land Exchange are the most important procedures from the surveyor's point of view. The implementation of the planning process, surveying of boundaries within the area of land consolidation, and the execution of changes in the cadastral map are performed by surveyors.

Law does not (yet) order the ecological aspects of Austrian agrarian reform. Indirectly, the farmers are motivated by financial subsidies granted to preserve existing landscape elements, to establish new landscape elements and field windbreaks, and to utilise more sustainable forms of cultivation to fulfil the demands of water retention and erosion control.

Land valuation

The taxation of rural land in Great Britain is essentially based upon profit, in other words, open market values. Agricultural valuation can be said to have started as a tenants right payment, when a tenant quite an agricultural holding. It is currently covered by the Agricultural Holdings Act (1986). The valuations are made on the basis of an open market value, taking into account the situation and character of the holding, as well as its productive capacity. Therefore, the valuation revolves around the yield, or potential yield, rather than the potential fertility of the soil. Local Valuers Associations normally fix consumer values for various produce and crops twice a year.

In Austria, the taxation of land is based on yearly averaged yield estimation. The main parameters for the land valuation are the natural yield conditions (such as soil quality, topography, climatic and water conditions) and economic yield conditions (traffic systems, together with the location and distribution of agricultural parcels). These factors are very similar to the considerations adopted by valuers in Great Britain. Soil specialists, employed by the Finance Authorities, estimate the value of land by using sample holes with a depth of 1m and by comparing it with sample standards. All matters connected with surveying, mapping and the Federal Office of Metrology and Surveying performs the parcel-wise evaluation of soil estimation. Parcel-related results of soil estimation are stored in the 'Real Estate Data Base' (GDB).

Interactions between cadastre, agrarian reform and land valuation

Figures 2 and 3 depict, as a schematic, the interactions between land registration, cadastre, agrarian reform and land valuation in Austria and Great Britain. Specific data flows, relating to surveying operations, are identified. The profession of experts and their interaction is also evident in the schematic.

In Austria, the Land Register (ownership register) and Parcel Register (cadastre) are, as mentioned above, in a common database (GDB). This is why the Land Register and Cadastre correspond in a perfect manner. Modifications concerning the registration of land are recorded by the Land Registry, whilst modifications to the Cadastre Register are based on partition maps, detailed by authorised surveying engineers and executed by regional (public) surveying offices. The owner of the parcel must pay the fees for the surveying engineers. For the registration of rights in the land register customers have to pay app. 1 per cent of the value of the object of contract and for the application to cadastre they pay small fee. The financing of the public authorities for the Cadastre and the Land Register are borne by the federal government.

In England and Wales, the Land Registry is a Government Executive Agency responsible to the Lord Chancellor. Its principal function is to register the title of land in England and Wales and to record all dealings once the land has been registered. A separate Land Register for Scotland is located in Scotland. The Land Registry is self-financing and has no calls on public funds. As mentioned above, the Land Registry employs no surveyors and is dependent upon maps produced by the Ordnance Survey. It maintains records, which are not directly referenced to those maps. Once a property has been registered, a small-scale plan (normally 1:2,500 or 1:1,250 but exceptionally larger) will be available of the parcel and an up-to-date and authoritative public record of ownership, rights, covenants and mortgages will be available.

The implementation of agrarian reform requires data relating to ownership and parcels. The Agrarian Reform Authorities of Austria can obtain this information free-of-charge from the Land Registry and the Surveying Offices. Experts of agrarian reform also using the results of the land valuation estimate the quality of soils within a consolidation project.

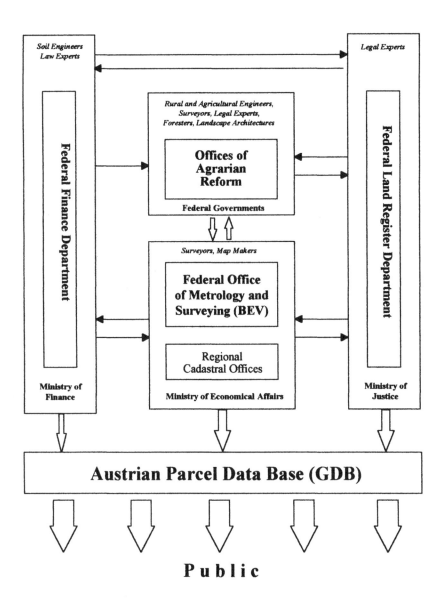

Figure 2 Data flow of parcel-based data in Austria

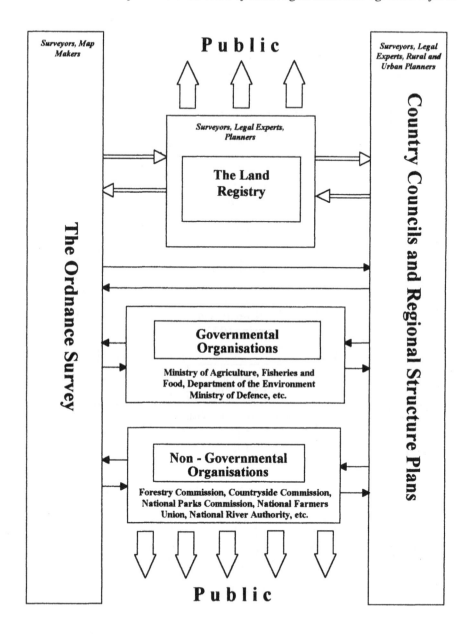

Figure 3 Data flow of parcel-based data in Great Britain

Surveying parcels, as well as the division of those parcels, can be performed - as regulated by Law (the Real Estate Division Law, §1) - by the Agrarian Reform Authority itself. After the completion of the land consolidation process (improvement of agricultural structures) the modified data are handed over to a special department of the surveying authority and to the Land Register, which revises the public registers. State Governments fund employees and the infrastructure of agrarian reform.

There are processes of data exchange and co-operations with the Finance Authority. Cadastral maps are required as the topographical basis for mapping of soil quality. The central surveying authority (the Federal Office of Metrology and Surveying), by linking the parcel geometry with the soil quality maps makes the evaluation of the parcel specific rateable value itself. With these data and information about the ownership of parcels, the finance authority finally calculates the personal taxes.

There is no comparable system in Great Britain.

Legal aspects

In Austria, laws and decrees regulate the administration, organisation and the implementation of Cadastre, Land Register, Agrarian Reform and Land Valuation. In one respect, this guarantees the stability of the system but, conversely, the regulations impose a degree of rigidity. The introduction of new aspects of scientific or administrative knowledge may be delayed before they can be established as working practice and integrated within a legal framework.

The most important Austrian laws concerning these subjects should be mentioned:

- Österreichisches Verfassungsgesetz (Austrian Constitution Law): Article 12 of the Austrian Constitution Law defines the measures of agrarian reform as stated below:

  1. land consolidation, division of agricultural communities, adjustment of common land use rights;
  2. adjustment and redemption of wood and grazing rights in forest land;
  3. land improvement by ways;

4. protection and improvement of alpine pastures;
5. the rural settlement sector (especially the relocation of farms).

The legal concept of agrarian reform exclusively refers to rural land problems. Agrarian reform is understood to be a national task, which is performed in co-ordination by the federal government and the federal states together. Principle legislation is the responsibility of the federal government, whilst legislation for implementation and the implementation itself are the responsibility of the federal states:

*   within the Grundbuchsgesetz (Land Register Law), the Grundbuchsanlegungsgesetz (Land Register Installing Law), and the Vermessungsgesetz (Surveying Law) the administration of the land register and the administration of the cadastre is regulated. The definition of surveying authorities and the practical realisation of land partition is part of the Liegenschaftsteilungsgesetz (Real Estate Division Law);
*   Flurverfassungs-Grundsatzgesetz (Land Consolidation Law) and Agrarian Reform Execution Laws (dependent on the Federal State): The Land Consolidation Law contains regulations concerning the procedure of land consolidation and land consolidation by voluntary land exchange, the division of agricultural communities and the adjustment of common land use rights. It further defines the responsibility of authorities, special procedural matters and determinations concerning land register and cadastre;
*   all affairs of land valuation are regulated in the Land Valuation Law.

The agrarian community of Great Britain is protected through a complicated web of law, policy and administration at the national and local level (Blunden & Curry 1988). The legislation varies between the countries of the United Kingdom, although the most significant differences occur in Scotland and Northern Ireland, which although sharing some of the countryside legislation with the rest of the UK, have largely different legislative and administrative systems. In a number of cases the regional legislation and administration tends to work in parallel with other regions.

The important laws related to England and Wales, apart from those outlined in Section 3, are:

- the Agricultural Act of 1947, which set out the principal objectives of agricultural support. The influence of the environmental character of the British and European countryside is largely a consequence of traditional agrarian practices. However, this particular Act was intended to increase food and agricultural production to ensure the availability of supplies and, moreover, to ensure that the supplies reached the consumers at reasonable prices. When the UK entered the EC in 1973, the responsibility for agriculture was largely transferred to the EC and had to conform to the aims for agriculture laid down in the Treaty of Rome, which were largely identical to the 1947 Act. To satisfy those aims, the agriculture industry across Europe has been essentially set on a collision course with the landscape conservationists and, at present, the pendulum is swinging in favour of conservation rather than production, as is mentioned above;

- the Town and Country Planning Act of 1947 was the first major act concerning the conservation of rural areas. In this act, the conservation of the countryside was to be achieved, in general, by preventing uncontrolled urban development. This Act established the framework for planning systems which still provide the principal means of controlling development in the open countryside;

- the National Parks and Access to the Countryside Act of 1949 gave the legal framework for the creation of National parks in England and Wales. It also defined the Areas of Outstanding Natural Beauty (AONB) and set up the National Parks Committee (NPC). This Act also set up the legal framework for establishment of National Nature Reserves (NNRs) and Sites of Special Scientific Interest (SSSIs). The situation in Scotland was somewhat different insomuch that the protection and conservation of the countryside was retained within the planning system. Concern during the early 1960s led to the introduction of the Countryside (Scotland) Act and the Countryside Act of 1968, which amended the original Act of 1949. The significant element concerning this amendment was an obligation for state users of natural resources (Ministry of Agriculture, Fisheries and Food, amongst others) to have a regard for nature conservation when carrying out their duties;

- the Town and Country Planning Act of 1990. This Act has, as its origin the Act of 1947, which was gradually expanded to accumulate a massive body of primary and subordinate legislation (Gregory, 1994). The Act of 1990 has some 337 sections and 17 schedules. Apart form the control of development, the Act provides legislation for environmental assessments, the protection of agricultural land, conservation by landowners and farm tenants, conservation of forests, trees and hedgerows, and of common land. It also reaffirms the significance of regional structure plans, which are particular emphasis in rural areas;
- the Planning and Compensation Act of 1991 strengthens the policy of regional structure plans, and states that planning decisions must accord with the structure plan unless material considerations dictate otherwise. As a result, strategies rural policies and countryside recreation have had to be developed for all parts of England and Wales. The National Park authorities, for example, have to prepare and implement management plans for their areas;
- the Environmental Planning and Policy Guide (PPG7) of 1992 concerns 'The Countryside and the Rural Economy'.

All the specified laws above are national laws and they do not address the issues of land registration and agrarian reform *per se*. In addition to this legislation, there are additional European and Global Guidelines and Ordinances that especially influence the procedures of agrarian reform in both countries. In particular, the numerous documents of the Agenda 21 (Conservation of Biological Diversity and Sustainable Use of Natural Resources of the Conference of Environment and Development), the Fauna-Flora-Habitat-Guidelines and Agenda 2000 of the European Union (1992), together with the decree Nr.2078/92 of the EU should be mentioned. Within the latter, the support for ecological and sustainable agricultural production methods is integrated. Since the ecological component of agrarian reform is not part of the national legislation of either country, decree Nr.2078/92 guarantee, through both national and European grants, the consideration of environmental and ecological measures within the land reform.

In both Austria and Great Britain, measures of agrarian reform are restricted to non-urban areas. However, within those rural regions the definition and designation of conservation sites can affect the execution of land reform. In the case of Austria these include:

- *National Parks*. In Austria, 5 National Parks are established yet covering about 3 percent of the country's area, with the major part in the alpine regions. Their implementation is the competence of the federal states whereas very often co-operation between several states is necessary. According to the national park laws agricultural land use is restricted to traditional extensive cultivation forms indirectly prohibiting measures of agrarian reform;

- *Nature Conservation Areas and Landscape Protection Areas*. The federal states have the option of implementing Nature Conservation Areas and Landscape Protection Areas by issuing decrees for the protection of valuable landscapes such as wetlands, lakes and their shores, forests, glaciers and alpine areas. In general all interference is prohibited, however, exceptions are possible (e.g. extensive agricultural land use). For several measures concerning agrarian reform like land melioration or rural infrastructure projects a permit of the federal states conservation authority is required. In practical planning work no measures of agrarian reform are carried out in such areas;

- *Water Conservation Areas and Flooding Areas*. These protective measures are defined by official decrees or orders of the Water Authority. They restrict land use and therefore impact upon many planning tasks of agrarian reform in a way that farmers are interested in exchanging the protected land with public land outside the designated area.

In England and Wales, designated sites and areas of conservation include:

- *National Parks.* The National Parks occupy approximately 9 per cent of the land area of England and Wales. The purpose of the 1949 Act was to preserve and enhance the natural beauty of the areas whilst promoting their enjoyment for the public. Planning authorities employ the policy and planning restraints and operate a modification of permitted developments which includes farming practices and the conservation of the countryside;

- *Areas of Outstanding Natural Beauty (AONBs).* These are covered by the same acts pertaining to the National Parks. They do not encourage public access and have more limited affects upon agrarian reform;

- *Sites of Special Scientific Interest (SSSIs).* These where defined loosely in the 1949 Act but were given more realistic status as protected sites in the Countryside Amendment Act of 1985;

- *National Nature Reserves (NNRs).* These where also defined loosely in the 1949 Act but have were given increased protection status in the Wildlife and Countryside Act of 1981;

- *Environmentally Sensitive Areas (ESAs).* These sites, designated by Ministers under the Agriculture Act of 1986, are an implementation of EC policy. They form a means by which a farmer may be contracted to adopt particular farming techniques, which promote conservation policies and practices.

The Environmental Agency was formed in 1996 to bring together the majority of Statutory Bodies relating to land use. Those former bodies included the National River Authorities, HM Inspectorate of Pollution, waste regulation authorities and many of the Department of the Environment's units. The principal aim was of consolidating the land use and environment of England and Wales.

The most significant difference between the legal systems of the two countries lies in Austrian Constitution where the Laws are more closely defined and easier to regulate. Due to the complexities of the laws of Great Britain, the use and abuse of rural land can fall under many Acts of Parliament, e.g. the various Highways Acts, the Finance Acts, etc.

In Austria the use of rural and urban land is regulated in several federal laws with spatial impact like the Federal Roads Law, the Water Law, the Forest Law etc. as well as in the Regional Planning Laws of the federal states. The latter contain standards on the general and specific goals of regional planning like the establishment of equal living conditions in all parts of the federal state, the development of infrastructure and housing providing a space-saving utilisation of land. The heart of the regional planning laws is the determination of planning instruments. This is reflected in the supra-local planning work carried out by the federal states in the form of comprehensive and sectored planning for the state territory and for regions. It is also evident in the planning work of municipalities (local development schemes, land use plan (zoning), building regulation plan) as well as the regulation of obligations and procedures. Agrarian reform measures are more or less affected by all those regulations. The most important affect on land consolidation lies within the local land use plans which designate settlement areas, building land and open land (including forests, agricultural and recreational areas) and therefore define those areas in which land consolidation can be carried out.

Future aspects

The importance of cadastre and land register is increasing. There are many new tasks and considerations that need information about land parcels and ownership. The monitoring and controlling of ecological measures (subsidised by national governments and/or the European Union) or Ecological and Environmental Impact Assessment (EIA) · within the planning period of infrastructure projects, are both based upon the knowledge given by these two systems together with additional attribute data.

The agrarian funding has changed within the last years: subsidies will now be based upon areas rather than gross yields. Planning data must therefore have the information of extent and ownership of parcels.

As the costs of land increase, the demarcation of parcels must be documented in an exact manner to avoid disputes. The boundary cadastre is the ideal method for this purpose. In both of the countries, many data sets with parcel-based information are available. Some of them are in a digital format, others exist in an analogue version as maps. In future all spatial data sets will have to be connected with each other and be integrated within a common database to known and defined standards.

The summary term for this parcel-based information is a Land Information System (LIS). This will be a tool for planning, monitoring and for controlling all spatially and temporally related processes of the land, including land registration, cadastre, agrarian reform and land valuation.

Conclusion

Within this paper, the systems of land registration, cadastre, agrarian reform and land valuation for both Austria and Great Britain were presented. Differences in historical evolution, in technical, organisational and in legal aspects were identified and considered. Because of the wide spectrum of tasks within the described field of activities, only central points of both systems could be considered.

A comparison between the relative merits of the Austrian and Great Britain systems has been consciously avoided: each system having developed under different historical frameworks. Similarly each system has a long tradition and its own evolution. Both systems work in their countries with few problems, and administrative and organisational structures were adjusted in an optimal manner. The two systems are also evolving to address the future needs and to take into consideration the current technological advances.

In the research leading to this paper, the authors have developed a hypothesis that there are more common elements than differences between the two systems currently in place. In both countries, there are general trends in agricultural production towards a change from commercial methods to ecologically controlled production. For this purpose there is a need to implement a common digital database for all the parcel-based information: LIS is the tool of the future to record, monitor and plan both land registration and agrarian reform.

References

Amtsblatt für das Vermessungswesen – Sondernummer 1981, *Das Vermessungsgesetz.* Herausgegeben vom Bundesamt für Eich- und Vermessungswesen, Jänner 1981.

Anhammer, G. (1980), *Das Verfahren der Grundstückszusammenlegung,* Manz - Verlag, Wien.

Austrian Federal Office of Metrology and Surveying (1997), *Report 1997 for "The National Mapping Agencies of Europe (CERCO).*

Blunden, J. & Curry, N. 1985. *The Changing Countryside,* Croom Helm, London.

Blunden, J. & Curry, N. 1988. *A Future for Our Countryside,* Basil Blackwell Ltd, Oxford.

Butlin, R.A., 1982. *The Transformation of Rural England c. 1580 – 1800: A Study of Historical Geography,* Oxford University Press, Oxford.

Dale, P.F., 1976. *Cadastral Surveys Within the Commonwealth,* HMSO, London.

Dwyer, J. & Hodge, I., 1996. *Countryside in Trust,* John Wiley and Sons, Chichester.

Gregory, M. 1994. *Conservation Law in the Countryside,* Tolley Publishing Company, Croydon.

Hochwartner, A. (1981), Grundstücksdatenbank - Agrarverfahren, in: *EVM - Eich- und Vermessungswesenmagazin,* Heft 33/81, Wien.

Hrbek, F., Zimmermann, E., Twaroch, Chr. (1985), The Austrian Parcel Data Base, in *Österreichische Zeitschrift für Vermessungswesen und Photogrammetrie,* 73.Jahrgang, Wien.

Kain, R.J.P., Baigent, E. (1992), *The Cadastral Map in the Service of the State,* The University of Chicago Press, Chicago 1992.

Kandutsch, G. (1995), *Die Grundbuchsentwicklung in Österreich,* Dissertation an der Karl-Franzens-Universität Graz.

Krammer, J. (1976), 'Analyse einer Ausbeutung I. Geschichte der Bauern in Österreich', in: *Sachen,* Heft 2, Wien.

Koenisberger, H.G., Mosse, G.L. and Bowler, G.Q. (1989), *Europe in the Sixteenth Century (2nd Edition),* Longman Group Limited, Harlow, England.

Kronsteiner, O. (1987), 'Probleme und Ziele der Agrarischen Operationen in Österreich', in: *Zeitschrift für Kulturtechnik und Flurbereinigung,* 28.Jhg., Heft 5, pp 310 - 317.

Lego, K. (1968), *Geschichte des Österreichischen Grundkatasters,* Bundesamt für Eich- und Vermessungswesen Wien, Eigenverlag.

Maland, D. (1983), *Europe in the Seventeenth Century (2nd Edition),* Macmillan, Basingstoke, England.

Northfield, 1979. *Committee into the Inquiryinto the Ownership and Occupation of Agricultural Land.* Cmnd 7599, HMSO, London.

Orwin, C.S. & Whetham, E.H., 1964. *History of British Agriculture 1846-1914.* Longmans, London.

Richeson, A.W. (1966), *English Land Measuring to 1800: Instruments and Practices,* The Society for the History of Technology and the M.I.T. Press, Cambridge (Mass.) and London (GB).

Riddall, J.G. (1983), *Introduction to Land Law (3rd Edition),* Butterworths, London, England.

Rowton Simpson, S. (1984), *Land Law and Registration (2nd Edition),* Surveyors Publications, London.

Thompson, F.M.L., 1963. *English Landed Society in the Nineteenth Century*, Routledge & Kegan Paul, London.

Trevelyan, G.M., 1988. *A Shortened History of England*, Penguin, London.

Wilflinger, J. (1973), Die Agrarischen Operationen in Vergangenheit und Zukunft, In: *Vorträge der Studienrichtung Kulturtechnik und Wasserwirtschaft zur 100-Jahrfeier der Hochschule für Bodenkultur*, Eigenverlag, Wien.

www.europa.eu: *Agenda 2000 of the European Commission*, Http://europa.eu.com.

www.ordsvy.gov.uk: OS Information Paper 7/1994: The national Land Information Service (NLIS) and other Related Developments.

Encoding expert opinion in Geo-Information Systems: a fuzzy set solution

Professor Allan J. BRIMICOMBE
School of Surveying, University of East London
Longbridge Road, Dagenham, Essex RM8 2AS

Abstract

In an increasingly Internet connected society, an inevitable consequence for land management and property professionals will be a shift in role from 'information brokers' to 'interpreters and managers of information'. Since much of the relevant information will be derived from digital spatial data, geo-information systems (GIS) and related IT will be key tools. An area where these tools currently perform poorly is in the encoding, storage and analytical handling of expert opinion or linguistic statements regarding the data content of GIS. This paper proposes a solution - fuzzy expectation ($\approx E$). Fuzzy expectation is derived from a small number of stylised fuzzy sets which, using an intuitive probability interface, are building blocks for 'translating' expert opinion into a compact fuzzy set representation. Once encoded, expert opinion about the data is embedded in the data structure and can be combined and propagated through GIS analyses such as overlay. A worked example in a land management context is provided as a means of illustrating the implementation of $\approx E$.

Introduction

Research findings of a survey on 'the Internet and the Property Profession' by The College of Estate Management (Dixon, 1998) show that the profession has much augmented its access to and use of IT. The most important use overall of the Web is 'research'. Despite substantial quantities of data and information being posted on the Web from which to conduct such research, respondents to the survey did not consider their role as 'information broker' under threat. Nor did they see the Internet impacting on location and land use decisions in the short term. Nevertheless, it is clear that the Internet is having a substantial impact on business and the professions. Commercial use of the Web is rapidly expanding in all sectors with 1995 figures suggesting about 1,500 businesses becoming newly connected each month as users are estimated to top 40 million and increasing by 10% a month (KPMG, 1997). Electronic commerce (*e*-commerce), that is, purchasing business-to-consumer and business-to-business over the Internet, is the new business frontier with turnover in Europe alone predicted to triple between 2000 and 2002 to ECU600 million and creating some 500,000 jobs related to the network economy (Condrinet, 1998).

A number of related initiatives in the UK are likely to have an impact on land management and the property profession sooner than respondents to The College of Estate Management survey may realise:

- the Law Commission has proposed electronic conveyancing as a means of speeding up home buying (Law Commission, 1998) and is currently being trialled by the Stroud and Swindon Building Society (HM Land Registry, 1998);
- the National Land Information Service (NLIS), designed to provide an on-line 'one stop shop' for land and property information, has completed trials in Bristol. This initiative is supported by others - the National Land and Property Gazetteer, the UK Standard Geographical Base and the National Topographic Database - all falling under the umbrella of the National Geospatial Data Framework (NGDF) (HM Land Registry, 1997);
- Geo-Information Systems (GIS) have an increasingly important role in spatial data integration and delivery to the public over the Web (MacEachren, 1998). GIS is likely to be a key technology in facilitating electronic property searches and conveyancing. Given the availability of high resolution data, such as Ordnance Survey's ADDRESS-POINT, GIS can now be used to extract and calculate the locational factors affecting property prices (Orford, 1998) whilst at other scales GIS-linked computational techniques such as cellular automata are being used to model land use change and urban growth (e.g. Clarke *et al.* 1997, Batty, 1998).

Nevertheless, respondents to the survey did consider that in the medium term, there would be, as a consequence of the Internet, an inevitable shift towards a role as 'interpreter' and 'manager' of information. It is this impending shift in their role that is central to this paper.

In a networked society based upon a largely anarchistic Internet, land and property data having a range of qualities (bad through good, low resolution through high resolution, highly generalised through highly specified) will be widely available either for free or at a cost to both layman and professional alike. Such a scenario sets up important opportunities for the professional (Brimicombe, 1998). The ability to integrate such data, evaluate quality and pertinence and provide interpretation to create actionable information will be a key attribute of the land manager and property professional. Since much of the relevant digital data will be spatial, or attributes of spatial objects, GIS and related IT will be key tools. As with any change, applied research is needed to explore the potential and limitations and to evolve the necessary tools.

The objective of this paper is to explore the impending issue, given the above scenario, of incorporating expert opinion or linguistic statements concerning the quality or pertinence of spatial data in GIS databases and to present a solution based on the use of fuzzy sets. The following section briefly reviews the relevant issues in GIS data modelling and spatial data quality. Fuzzy sets as a traditional means of encoding linguistic hedges are introduced and a critique of their shortcomings developed. This sets the context in which an enhanced technique, *fuzzy expectation* ($\approx$E) is developed and presented. A worked example in a land management context is then provided to illustrate the implementation of $\approx$E.

Geo-spatial semantics and expert opinion

The concept of 'spatiality' in land management can be viewed as a semantic sequence as illustrated in Figure 1. Space, at its most objective, is a mathematical space of co-ordinate geometry. The primitive elements are *point, line* and *polygon* or *cell*. At a socio-economic level, the neutrality of space is replaced by an intrinsic superiority or inferiority for some purpose.

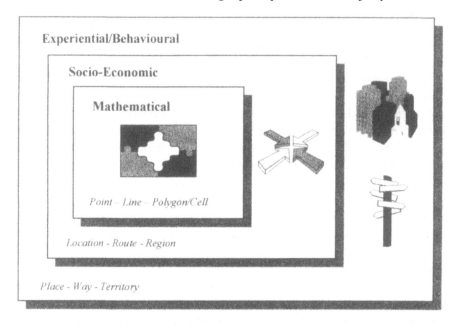

Figure 1   The geo-spatial semantic and its primitive elements (developed from Couclelis, 1992)

A location has both site and situation as a result of spatial relations between consumers, producers, labour and raw materials. Here the primitive elements are translated into *locations*, *routes* and *regions*. Finally, in the experiential/behavioural domain, the spatial primitives are further translated through an infusion of human meaning into *place*, *way* and *territory*. Thus in any study of significant land issues involving GIS, the data adopts a range of semantics. In moving from GIS to decision making, the spatial elements necessarily take on a growing imprecision and the user increasingly relies on interpretation of the data and its meaning through spatial reasoning. However, during the initial data collection and encoding for GIS, most of the spatial interpretation associated with place is stripped away and the real-world fluidity of the spatial elements are reduced leaving the user to somehow re-establish these during analysis and decision making. New methodologies are required that allow data to retain a maximum of these interpretative elements and fuzziness so as to increase their application relevance and yet can be efficiently handled by computational modelling. One approach would be to encode and embed within the geometric representation of homogeneous class intervals (e.g. land use classes) expert opinion on the degree to which these fit an adequate and useful representation of reality.

Fuzzy sets and their use in a spatial context

Fuzzy sets (Zadeh, 1965; Kaufmann, 1973) were developed to handle the imprecision inherent in much of human reasoning (Figure 2).

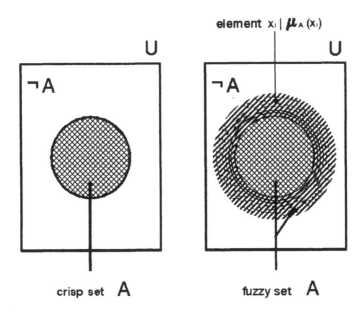

Figure 2 An illustration of crisp and fuzzy sets

A fuzzy set assigns levels of membership μ in a range [0, 1] for each element of x in a set A in a universe U:

$$\forall x \in U, \ \{x \mid \mu_A(x)\} \ ; 0 \ge \mu_A(x) \le 1 \tag{1}$$

Hence for intervals of x of 0.1 in the range [0, 1]:

$$A = \{0|\mu_0, \ 0.1|\mu_{0.1}, \ 0.2|\mu_{0.2}, \ 0.3|\mu_{0.3}, \ \ldots\ldots\ldots \ 0.8|\mu_{0.8}, \ 0.9|\mu_{0.9}, \ 1|\mu_1\} \tag{2}$$

The traditional binary (0, 1) are 'crisp' and can be viewed in set form:

$$\neg\mathbf{A} = \{0|1\} \qquad \mathbf{A} = \{1|1\} \tag{3}$$

(where $\neg$ is "not") and can hence be viewed as special cases of a fuzzy set. Fuzzy sets can be combined in Boolean operations and in general can be handled in much the same way as probabilities:

Intersection ($\cap$) **A** AND **B**: $\forall x \in \mathbf{U}, \ \mu_{A\cap B}(x) = \text{MIN}\,(\mu_A(x), \mu_B(x))$ (4)

Union ($\cup$) **A** OR **B**: $\forall x \in \mathbf{U}, \ \mu_{A\cup B}(x) = \text{MAX}\,(\mu_A(x), \mu_B(x))$ (5)

Fuzzy set operations and their use in geography and the spatial sciences, have been reviewed by Macmillan (1995). Gale (1972, p341), in an early references to fuzzy sets in a geographical context, characterises three forms of 'inexactness' arising in classification due to:

- insufficient information about an object or area;
- the neutrality of predicates, that is, it is not clear into which class an object falls;
- secondary effects arising from the nature of the observation or observer (e.g. obstructed view or lack of access).

The use of fuzzy sets in resolving such class membership problems has been studied by Leung (1984, 1987), Burrough *et al.* (1992) and extensively in the context of remote sensing (e.g. Goodchild, 1994; Foody, 1995). In this paper, emphasis is placed on the subjective conviction of an expert to place an object within a class arising out of insufficient evidence, the presence of heterogeneity or secondary effects either individually or in combination.

The term 'fuzzy' has been increasingly use of in the context of GIS (e.g. Kollias & Voliotis, 1991; Burrough *et al.*, 1992; Sui, 1992) but has been applied loosely by some researchers to any non-binary treatment of data such as probabilities. Probabilities, however, remain crisp numbers despite giving greater discrimination in the range [0, 1] and should not be confused with fuzzy sets. The use of fuzzy set theory proper in GIS has thus far been quite restricted (Unwin, 1995) and is reviewed by Altman (1994). One area of application has been to quantify verbal assessments of data quality from image interpreters and as a consequence of expert evaluations (Hadipriono *et al.*, 1991; Sui, 1992; Gopal & Woodcock, 1994). These studies were broadly based around constructing linguistic scales between extremes (e.g. from 'absolutely wrong' to 'absolutely right' or 'least valuable' to 'most valuable') which were represented by fuzzy sets. Expert evaluations were expected to be restricted to the relevant linguistic scale.

The problem of fuzzy set encoding of expert opinion

Table 1 provides an example of a set of qualifying terms for features and boundaries. They were defined and encouraged for use by the Geological Society Working Party on Land Surface Evaluation for Engineering Practice (Edwards *et al.*, 1982) and are intended as qualitative indicators of data reliability. The problem that arises, however, is that such 'standard' terms are themselves defined by a further vocabulary which, to each individual, may have nuances and different interpretations.

	Qualifying term	**Definition**
Features	*Certain*	*Well defined, identifiable*
	Reliable	*Poorly defined, identifiable*
	Unreliable	*Deduced*
Boundaries	*Well defined*	*Full boundary distinct*
	Poorly defined	*Boundary mainly distinct*
	Partly defined	*Boundary mainly inferred*
	Estimated	*Boundary inferred*

Table 1 An example of terms suggested for use in qualifying mapped features and boundaries on the basis of expert opinion (from Edwards *et al.*, 1982)

One of the earliest and continuing applications of fuzzy sets has been to represent qualifying terms and to facilitate computing with words (Zadeh, 1972, 1996; Lakoff, 1973). Examples of such representation (from Ayyub & McCuen, 1987) are:

Small, low, short or *poor*: $\mathbf{A} = \{ 0|1, 0.1|0.9, 0.2|0.5 \}$ (6)

Medium or *fair*: $\mathbf{A} = \{ 0.3|0.2, 0.4|0.8, 0.5|1, 0.6|0.8, 0.7|0.2 \}$ (7)

Large, high, long or *good*: $\mathbf{A} = \{ 0.8|0.5, 0.9|0.9, 1|1 \}$ (8)

These are illustrated graphically in Figure 3. They reflect an empirically verified general pattern whereby fuzzy sets have reduced spread as the qualifying terms verge on the extremes equivalent to binary 0 or 1. These are in contrast to the arbitrarily defined terms and fuzzy set representations often found in the literature (Figure 4).

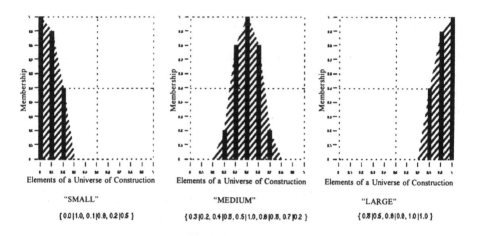

"SMALL"

{ 0.0|1.0, 0.1|0.8, 0.2|0.5 }

"MEDIUM"

{ 0.3|0.2, 0.4|0.5, 0.5|1.0, 0.8|0.8, 0.7|0.2 }

"LARGE"

{ 0.8|0.5, 0.9|0.8, 1.0|1.0 }

Figure 3 Illustrative representation of fuzzy sets (6) - (8)

Whilst the construction of a fuzzy set to represent expert opinion may appear to be a fairly straightforward (if not arbitrary) process, the same cannot be said for the reverse. Faced with a fuzzy set, which does not appear to fit any pre-defined terms, it can be very difficult to 'translate' it into an appropriate linguistic. Furthermore, fuzzy sets are cumbersome to store in a database. Not only is the notation difficult to encode but there are 39,916,789 useful combinations of fuzzy sets in the range [0, 1] for an interval of x_i=0.1 These are serious problems in the use of fuzzy sets. Whilst they may at first appear to be an attractive solution (commented on by many authors), they have not yet found widespread application in GIS.

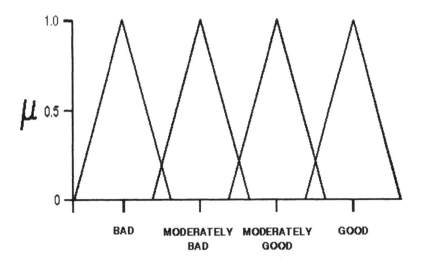

BAD   MODERATELY   MODERATELY   GOOD
          BAD   GOOD

Figure 4   An example of arbitrarily defined, equally spaced fuzzy sets that do not conform with empirically verified patterns

From fuzzy sets to fuzzy expectation, $\approx E$

To overcome these practical problems in the use of fuzzy sets, a universal 'translator' has been devised to assist in the two-way translation of expert opinion and fuzzy sets (Figure 5). This translator comprises of two parts: an array of 11 stylised fuzzy sets spanning the range [0, 1] coupled with a corresponding array of 11 intuitive probabilities. The central portion of Figure 5 gives an array of fuzzy sets described by values of μ_A for intervals of x_i=0.1 from 0 to 1. Blanks are assumed to have values of $\mu_A(x_i)$=0. Two of these fuzzy sets are displayed graphically to aid visualisation. These 11 fuzzy sets are stylised and have a number of attributes that make them particularly useful as linguistic building blocks:

- the stylised set begins and ends with binary (crisp) numbers 0 = {0|1} and 1 = {1|1};
- the nine intermediate stylised fuzzy sets are spaced with their maximum membership of $\mu_A(x_i)$=1 stepped across the x_i range at 0.1 interval. Thus each fuzzy set has its maximum membership

uniquely placed within the range;

- the roughly triangular form of the stylised fuzzy sets spreads towards the intermediate, mid-range of the stylised fuzzy sets. Thus where there is greatest uncertainty (mid-way in the [0,1] range), the stylised fuzzy sets are most spread to reflect higher levels of uncertainty. The reduced spread of the fuzzy sets as they approach binary 0 or 1 accords with the empirical evidence for fuzzy set representations of linguistic hedges;

- the Hamming (orthogonal) distance between each of the stylised fuzzy set and its immediate neighbour is constant at 2.00 indicating that the stylised fuzzy sets unambiguously partition up the space over the range [0, 1]. The Hamming distance between two fuzzy sets is calculated by:

$$d(\mathbf{A}, \mathbf{B}) = \sum_{i=1}^{n} |\mu_A(x_i) - \mu_B(x_i)| \qquad (9)$$

The interface to the array of stylised fuzzy sets is through a simple array of intuitive probabilities. A common form of hedge, other than purely linguistic, is intuitive (subjective) probabilities which individuals use in making judgements under uncertainty (Tversky & Kahneman, 1974). Thus an individual may say "I'm 80% sure" though this is not meant in itself to be a precise evaluation but a broad reflection of the degree of certainty. Since both types of hedges - linguistic and intuitive probability - are frequently used, it is possible for an individual to make an equivalence between the two. Thus "I'm reasonably sure" may, for an individual, have an approximate equivalence to "I'm 80% sure". This would be for each individual to define for their own range of vocabulary. Given the way the stylised fuzzy sets in Figure 5 step across the range [0, 1], each stylised fuzzy set can be 'labelled' or identified by the x_i where $\mu_A(x_i)=1$, that is, its point of maximum membership. These labels at 0.1 interval in the range [0, 1] provide an intuitive probability-like metric which is named here as *fuzzy expectation* ($\approx$E). Thus the choice of a value of $\approx$E as an intuitive probability, delivers an underlying stylised fuzzy sets to express a level of uncertainty. An expert opinion 'translated' into its equivalence of one or more intuitive probabilities (value(s) of $\approx$E) has the underlying fuzzy sets, combined using a Boolean OR, substituted for it.

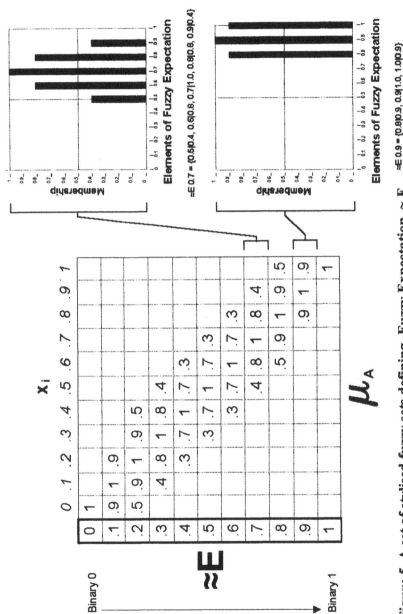

Figure 5 A set of stylised fuzzy sets defining Fuzzy Expectation, ≈ E

The reverse process is also available where fuzzy sets can be translated into statements of fitness-for-use as illustrated in the worked example below.

The problem of encoding and storing numerous cumbersome fuzzy sets against each feature in a GIS database is now addressed. To overcome this, the values of ≈E representing an expert opinion are stored within the data structure as integer pointers to a lookup table containing the 11 stylised fuzzy sets. Each polygon object in a GIS has a label (e.g. 'residential'). In a 32 or 64 bit architecture, there is usually ample room to store these integer pointers by placing them within the spare bits of such a label using a C language struct and thus have no adverse effect on database size. Hence there is a simple mechanism of embedding the expert opinion within the datum to which it relates.

An example of using fuzzy expectation

Consider a scenario where for an urban regeneration study, spatial information on functional zones (industrial, retail, residential) needs to be mapped against zones representing the condition of existing building stock. Clearly these two data layers require an element of judgement in their construction and the zones themselves may be more or less heterogeneous as a consequence of generalisation. The compiler of each layer can express his/her uncertainty in the labelling of a zone through a linguistic opinion (Figure 6). On entering the data into the GIS, the compiler of that layer defines an equivalence for their vocabulary of opinions in terms of ≈E. In doing so, the compiler only has to match intuitive probabilities or sets of intuitive probabilities to the linguistic opinions and pointers to the underlying stylised fuzzy set(s) are substituted into the database. Users of the system do not have to think in terms of fuzzy sets, only in terms of their own vocabulary and its equivalent intuitive probabilities. A number of points can be noted about this data entry process:

- a linguistic may be equivalent to one or more values of ≈E;
- these values would normally be adjacent in the series (logically) but need not necessarily be so;
- linguistic opinions can overlap in their ≈E equivalence showing that two linguistic terms may be close in meaning;
- where two or more values of ≈E are used, they are not used

singly but are combined using a Boolean OR prior to propagation through analysis;

- a table giving the expert's vocabulary and its $\approx$E equivalences should be stored as metadata for future reference.

It must be stressed that the process illustrated in Figure 6 is not proposing a fuzzy logic alternative to standard GIS analyses using Boolean logic. $\approx$E is a means of encoding, propagating and decoding linguistically expressed uncertainty associated with spatial objects. $\approx$E is designed for use with standard GIS functionality to give a quality statement of support for the outputs. Naturally, a data user's linguistic expression of fitness-for-use will be different from the data compiler's linguistic statements of uncertainty. But the fitness-for-use statements such as "highly applicable" or "useless" are still qualifying statements to the same degree that the data compiler's expert opinions are and therefore the approach is the same. In the example given in Figure 6, linguistic opinions encoded as fuzzy sets through $\approx$E have been propagated through an intersection overlay of the two map layers to give a resultant fuzzy set whose 'meaning' may not be known. To match this propagated uncertainty with the relevant expression of fitness-for-use (in terms of $\approx$E), the resultant fuzzy set needs to be equated with one of the stylised fuzzy sets and thus with one of the user's equivalent quality statements. This is achieved by calculating the relative Hamming distance between the resultant fuzzy set and all the stylised fuzzy sets in $\approx$E:

$$\delta(\mathbf{A},\mathbf{B}) = \left\{ \sum_{i=1}^{n} \left| \mu_A(x_i) - \mu_B(x_i) \right| \right\} \Big/ n \qquad (10)$$

The shortest distance provides the match and hence the resultant fuzzy set can be 'translated' into one of the user's linguistic criteria. $\approx$E and its underlying fuzzy sets permit expert opinions of uncertainty to be stored, propagated through analyses and output as linguistic expressions of fitness-for-use.

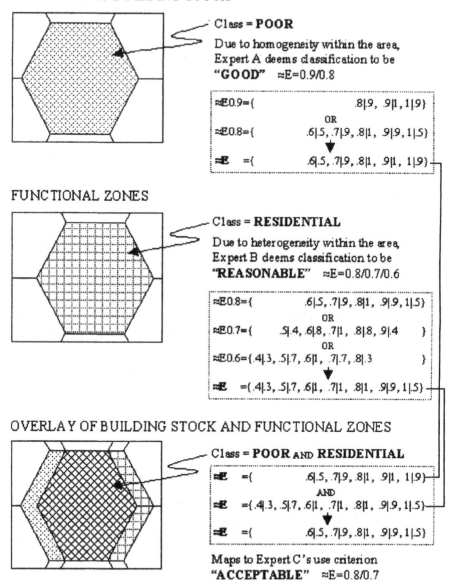

Figure 6 An example of propagating expert opinions on data quality into a statement of fitness-for-use during a GIS overlay operation

Conclusions

A new technique has been devised that permits an objective handling of expert opinions of data reliability specifically within GIS and more generally within the geo-spatial semantic. By using fuzzy expectation as building blocks for translating expert opinions into stylised fuzzy sets, the incorporation of interpretative or contextual human meaning into a mathematical conceptualisation of space becomes possible. Many of the difficulties in using fuzzy set descriptors in GIS and other technologies are overcome. The expert opinions embedded as stylised fuzzy sets can be propagated using Boolean operators to give a resultant fuzzy set which can be 'translated' back into a linguistic fitness-for-use statement. Thus for the first time, linguistic criteria of fitness-for-use can be derived for the outcomes of GIS analyses.

References

Altman, D., 1994. Fuzzy set theoretic approaches for handling imprecision in spatial analysis, *International Journal of Geographical Information Systems*, **8**, 271-289.

Ayyub, B. & McCuen, R., 1987. Quality and uncertainty assessment of wildlife habitat with fuzzy sets, *Journal of Water Resources Planning and Management*, **113**, 95-109.

Batty, M., 1998. Urban evolution on the desktop: simulation with the use of extended cellular automata, *Environment & Planning A*, **30**, 1943-1967.

Brimicombe, A.J., 1998. The future of land surveying education: meeting the needs of a changing profession, *Surveying World*, **6**(4), 23-26.

Burrough, P., MacMillan, R., & van Deursen, W., 1992. Fuzzy classification methods for determining land suitability from soil profile observations and topography *Journal of Soil Science*, **43**,193-210.

Clarke, K.C., Hopper S. & Gaydos, L., 1997. A self-modifying cellular automation model of historical urbanisation in the San Francisco Bay area, *Environment & Planning A* , **24**, 247-261.

Condrinet, 1998. Content and commerce driver strategies in global networks, *http://www2.echo.lu/condrinet/* as viewed 27th November 1998.

Couclelis, H., 1992. Location, place, region and space, in *Geography's Inner Worlds* (eds Abler *et al.*), Rutgers University Press, New Jersey: 215-233.

Dixon, T.J., 1998. *Building the Web: The Internet and the Property Profession.* The College of Estate Management, Reading.

Edwards, R., Brunsden, D. Burton, A., Dowling, J., Greenwood, J., Kelly, J., King,

R., Mitchell, C., & Sherwood, D., 1982. Land surface evaluation for engineering practice, *Quarterly Journal of Engineering Geology*, **15**, 265-316.

Foody, G., 1995. Approaches for the production and evaluation of fuzzy land cover classifications from remotely sensed data, *International Journal of Remote Sensing*, **17**, 1317-1340.

Gale, S., 1972. Inexactness, fuzzy sets and the foundations of behavioural geography, *Geographical Analysis*, **4**, 337-349.

Goodchild, M., 1994. Integrating GIS and remote sensing for vegetation analysis and modelling: methodological issues, *Journal of Vegetation Science*, **5**, 65-74.

Gopal, S., & Woodcock, C., 1994. Theory and methods for accuracy assessment of thematic maps using fuzzy sets, *Photogrammetric Engineering and Remote Sensing*, **60**, 181-188.

Hadipriono, F., Lyon, J., & Thomas, L., 1991. Expert opinion in satellite data interpretation, *Photogrammetric Engineering and Remote Sensing*, **57**, 75-78.

HM Land Registry, 1997. *National Land Information Service Feasibility Study, Executive Summary*. HM Land Registry, London.

HM Land Registry, 1998. Press Notice 19/98: Electronic system for discharging mortgages to cut delays in house buying. *http://www.dmercer.mcmail.com/ lrp1998.htm* as viewed 5th January 1999.

Kaufmann, A., 1973. *Introduction a la théorie des sous-ensembles flous, Tome I: éléments théoriques de base*, Masson, Paris.

Kollias, V., & Voliotis, A., 1991. Fuzzy reasoning in the development of geographical information systems, *International Journal of Geographical Information Systems*, **5**, 209-223.

KPMG, 1997. *The Internet: A Guide to Business Users*, KPMG Management Consulting, London.

Lakoff, G., 1973. Hedges: a study in meaning criteria and the logic of fuzzy concepts, *Journal of Philosophical Logic*, **2**, 458-508.

Law Commission, 1998. Land registration for the twenty-first century: a consultative document, *Report of the Law Commission No. 254*. The Stationary Office, London.

Leung, Y., 1984. Towards a flexible framework for regionalisation, *Environment & Planning A*, **16**, 1613-1632.

Leung, Y., 1987. On the imprecision of boundaries, *Geographical Analysis*, **19**, 125-151.

MacEachren, A.M., 1998. Cartography, GIS and the world wide web, *Progress in Human Geography*, **22**, 575-585.

Macmillan, W., 1995. Modelling: fuzziness revisited, *Progress in Human Geography*, **19**, 404-413.

Orford, S., 1998. Valuing location in an urban housing market. *http://www.geog. port.ac.uk/geocomp/geo98/78/gc_78.htm* as viewed 2nd December 1998.

Sui, D., 1992. A fuzzy GIS modeling approach for urban land evaluation,

Computers, Environment and Urban Systems, **16**, 101-115.

Tversky, A., & Kahneman, K, 1974. Judgement under uncertainty: heuristics and biases, *Science,* **185**, 1124-1131.

Unwin, D., 1995. Geographical information systems and the problem of 'error and uncertainty', *Progress in Human Geography,* **19**, 549-558.

Zadeh, L., 1965. Fuzzy sets, *Information and Control,* **8**, 338-353.

Zadeh, L., 1972. A fuzzy-set-theoretical interpretation of linguistic hedges, *Journal of Cybernetics,* **2**(3), 4-34.

Zadeh, L., 1996. Fuzzy logic = computing with words, *IEEE Transactions on Fuzzy Systems,* **4**, 103-111.

Informal land delivery and management: a case of two housing areas in Zanzibar town

Marjukka VEIJALAINEN
Helsinki University of Technology, Helsinki

Abstract

This paper presents some of the findings based on a research project, which has been carried out in two, recently developed, unplanned settlements in Zanzibar town and in several public offices. The research was conducted between March and May 1996 and will continue in Mbeya town. The study was focussed upon defining the effect of informal land management practices to the housing in unplanned areas in Zanzibar town.

The study areas are within the Mto Upepo river valley and the surrounding area of Kibweni. The definition of town boundaries varies. However, both study areas are clearly part of the town although municipal services are not provided. Thus, both areas are outside the municipal boundary of Zanzibar town but inside the Master Plan (1982) boundaries.

This paper identifies the impact of such developments in Zanzibar, which results in a shift of interest from the organisation of land management within government bodies to those working to support the actions that should take place.

Introduction

In Zanzibar town the local people have developed the majority of the housing areas within the local community[1]. Typical of the cities in many developing countries, this so-called unplanned sector is expected to dominate the housing development in many towns (NLUP_S.01, 1995). Conventional planning and land management has had a minor role in the development of urban housing and the result has, in the case of several large projects, been unsustainable (Andriananjanirana-Ruphin, 1995; Myers, 1993).

At this current time, along with the political and economical transition, the official land use planning and land management systems are under reform in Zanzibar. The island-state is being integrating within the global economy (World Bank, Habitat conventions, etc.), and is expected to require more effective land management. Many African land management specialists are proposing that the realistic land management should, instead of copying the tools from developed countries, take into consideration the real circumstances pertaining in each country (Ochieng'-Akatch, 1994). Instead of empowering the experts to rule the urban development, Kironde (1995) has proposed co-operation between the formal and informal sector on land delivery issues. The South Saharan African Governments have been advised to formulate their land policies so that they take into account

> the informal sector, self-reliance and the needs of the urban poor (Symposium on Shelter, 1991).

This is one of the aims of the Zanzibar government. It has accepted a National Land Use Plan (NLUP_S.01, 1995) proposal

> to formalise the informal land market that exists and to accommodate greater private sector involvement in land use planning.

This statement is only the beginning of the assessment and adjustment work needed on administrative, legal and practical levels of land management. The greatest difficulty lies in the definition and, especially, in the implementation of a sustainable and working system of such co-operation. Unfortunately there are no available models of international local land management, partly due to the unique nature of land relations and socio-economic structures in each location.

In Zanzibar land legislation has faced many changes during the last few years (Törhönen, 1997). Unfortunately, the unplanned urban settlements have been treated only as a secondary problem and only a few parts of the legislation bring any light to the specific land problems in unplanned areas. The informal system of selling developments is not under any land management legislation. Law (Land Tenure Act, 1992) denies peasants the right to sell or subdivide their rural holdings (three-acre plots registered as rural holdings).

In practise, at the Commission for Lands and Environment (COLE), the Department of Surveys and Urban Planning (DoSUP) worked with a formalisation and planning experiment for urban informal housing area. In the latter, the planning was an expert activity, but the problem of land compensation[2] was solved by leaving the majority of the surveyed plots to the (rural) landowners (Muhajir, 1996). The project was carried out during 1995-1996 and it is not been duplicated since (Muhajir, 1997).

Internationally, the informal land development processes have been studied from diverse points of view (see Payne, 1989; Mabogunje, 1990; Carroll, 1980; Angel *et al.*, 1982; Falleth, 1993; Larbi, 1995; etc.). Until recent years, little has been written of the effects of different informal planning and land management actions on the outcome of informal development process (especially in secondary towns). Earlier works often concentrated to the physical pattern, land value development and living conditions in these settlements. Information on local land management processes is essential if informal land markets are to be legalised.

Among the few studies on the subject, Falleth (1993) has investigated the land actions and the development process in Katmandu and Patan in Nepal. Larbi (1995) has compared urban land development processes with the land policy in Ghana. In Tanzania the informal land delivery, together with the later development of the areas, has been the subject of interest as well. Kombe (1992) has made case studies on self-initiated development projects in informal housing areas. He also made a research on the formal and informal land management in Dar es Salaam (1996). Burra (1996) has reported one case study of autonomous, self-initiated planning and local spatial management for an informal settlement.

The purpose of this paper has been to increase the knowledge of the local management of land in growing, unplanned settlements in Zanzibar town. The hypothesis of Yahya & Nzioki (1994)

> that the government housing strategies for low income groups are rarely effective due to lack of understanding of and accurate information on the operations of housing sub-markets among the low-income groups in urban areas, especially in unplanned settlements

fits the process of land management as well. The transformation of agricultural land to urban housing plots is especially interesting and the informal land delivery is illustrated from the inhabitant's point of view. The major problems related to the process and experienced by the new urban dwellers, are described and discussed, together with the response of the people to the idea of land use planning.

Methods

The field study commenced with discussions and interviews between different government and municipal land development and management experts. At the next stage, local leaders expressed their views on the housing in their area of jurisdiction (semi-structured discussions). Rural landholders and local leaders were found to be valuable key informants, who were able to describe the local land management processes. Small groups of men and women from two selected areas were interviewed (semi-structured discussions). Other inhabitants of the two areas gave information and opinions on the observed 'planned' sites and on observed problems, both in the discussions and questionnaires. The questionnaires were presented in each house to the person who had acquired the site for dwellings. Both qualitative and quantitative data and conclusions could be drawn from the results of the questionnaire and the interviews. We also visited some other growing areas in order to certify that the processes found were not strictly unique to the study areas.

Findings

The formation of unplanned areas in Zanzibar town and the development process

During the last 15 years, land officials have been able to produce, on average, 11% of the applied plots in Zanzibar town. In the 1990s, the supply of plots has been more effective. For instance, in 1995 58% of the applied 990 plots were delivered (Source: COLE, DoSUP Statistics, 1996). How the number of plot applications is related to the needs of plots is never studied. The official control over the land development and building has been weak. Unplanned settlements are frequently growing, especially in the northern and eastern rim of the town.

The informal plot delivery takes place on undeveloped land. Private developers are few on the housing market (Group 5, 1992; Myers, 1993; Khadija, 1996). The main producers of dwellings are the people themselves.

126

The main type of dwelling is a one family house. Unplanned housing areas in Zanzibar are not slums; the water and electricity service providers do not discriminate between the areas (Saleh, 1996). The rate of owner occupancy is high.

The squatters or informal land delivery has been mentioned in many studies on Zanzibar. On the basis of the case studies, Group5 (1992) classified formal, semi-formal and informal ways to acquire a house or a plot in the town. Singer (1993) studied squatter settlements and land disputes in the town in 1993. He, however, did not record the commercial side of the process. Myers (1993) made a study on 'Reconstructing Ng'ambo: Town Planning and the Development of the Other Side of Zanzibar'. He (Myers, 1993; 1994) found that power 'uwezo', interrelated customary beliefs (Islamic law) and practises (neighbourliness and construction methods) have been the main forces to design the urban unplanned housing areas in old Ng'ambo. He also noted that among the reasons for partial failures of the colonial and post-colonial, large housing projects was the lack of interaction between the project planners and the people whose needs were to be satisfied. This study supports almost all findings of the above mentioned. The practise of asking the neighbours for a permit to settle (no compensation), defined by Singer (1993), was not found as such in the two study areas.

The land allocation process in the two study areas was highly commercial (75%). There were only a few cases where land was donated or the user of land had the right to use someone else's land. The free allocation of land to a (close) kin came up often in discussions. Each rural landholder is said to have either allocated 1 or 2 plots to a close relative (sister or a child) or reserved land for them. We found one separate housing group in Mwanyanya (next to Kibweni) where a resident claimed that this particular three-acre plot was reserved for the relatives of its holder only.

The land allocation process for housing plots is basically simple and fast in unplanned areas. The transaction needs little explanation. At best it takes only few hours to go through. The shortest process includes looking for the owner, negotiating, making an agreement paper and paying the compensation. The item to change hands may be trees, cultivations, foundations, huts or also, in few cases, simply the land. The trade of development is due to the common knowledge that the sale of three-acre plots and government land is prohibited. Same methods of land sale are used also on the mainland and in rural Zanzibar. The compensation of a plot may include payments to several people (holder of land, owner of cultivations, etc.).

The land allocation process is the most important element to the development of the structure of the housing areas. At this stage the planning is imperative although many land use decisions take place after the land allocation. Later negotiations, ignoring boundaries and solved disputes, impact the land use as well. The allocated plots, however, sets the rough limits to later land use decisions, especially if the plots are small and close to each other, as is often the case.

The relationship of the actors in the land allocation process

The basic actors in land allocation are the allottee, the allocator, the middleman, the local leader, witnesses and neighbours. People referred in this study as allocators are those who have subdivided and sold housing plots in unplanned neighbourhoods. The allottees are those who receive the plot.

The landholders and owners of the developments are the principal actors in informal areas. They have the power to plan, implement and allocate the plots. In practise they decide the location, dimensions and price of the plots. A few landholders were willing to enter negotiations about these issues. The role of the allottee becomes important after the allocation.

The relationship between the allocator and allottee is not very close, which may have some impact to the willingness to negotiate. In the Zanzibar study only 16% and 13%, respectively, of the respondents said that they are relatives or friends with the allocator of the land. The role of the relatives and friends as informants is more important when the salient house builder seek vacant plots (42%). The majority obtained the information and bought the land from a stranger. The sale of plots was gradual, on a plot by plot bases in most of reported cases and the sale of several plots at a time was also reported.

Once the allocation is in the hands of the allottees, the allocators reported that they are not interested in controlling the later development of the area. This would probably be difficult, since the property is sold and the allottees and allocators are seldom close to each other. If the allottee further sells the plot, the newcomer does not necessarily consider the ex-owner to have any right to control land uses.

The neighbourhood relations and nets are created gradually, but common problems seem to ease the co-operation with neighbours. In the study areas several people had negotiated with the neighbours on land uses or organised groups to improve the living environment[3]. Co-operation with neighbours may be quite successful during the installation of the infrastructure. For instance, 50% of water tap owners in Kibweni had discussed the acquisition of the connection with the neighbour. 75% of people in Mto Upepo valley were against the claim that one does not discuss the developments on ones own land with neighbours. In Kibweni 60% were against it. Naturally some people are affluent enough to act alone and the first-comers had practically no neighbours to negotiate with.

The middlemen and local leaders claimed that they are not involved in the definition of the dimensions and of the location of the land to be allocated. Later the role of local leaders, Shehas, as organisers of neighbourhood developments, becomes important in the settlement of disputes.

Traditionally land has been in the hands of the men. We were informed of only one female allocator (outside the study area). She had inherited a three-acre plot, because all the male relatives of her husband had recently obtained their own plots. Women discussing this matter considered her to be very lucky. In Kibweni one woman was reported to have 'given a three-acre plot back to government' after her husband had passed away. These lands were later subdivided and sold by a local leader. Most of the plot and house owners were men.

Planning

A land use plan is defined in this work as 'an orderly arrangement of land uses'. Drawings are not considered to indicate planning, but

> the plan always implies mental formulation and sometimes graphic representation (Merriam-Webster, 1997).

In unplanned areas the planning of land uses have different actors, tools, time spans and areas of implementation than the official planning. In unplanned areas no drawings are made on papers; however, this may come into question during the house-planning phase.

The land use plan is not presented for acceptance to upper authorities, but negations may take place between the allocator, allottee and other parties; the land of the allocator sets the limits for the planning area. Planning is implemented through land allocation; it may either be defined during the process or may guide it. The negotiations between neighbours, solutions of dispute and individual use of land further shapes the areas. Further planning and negotiations are required to install infrastructure and services.

Land use planning is comprehensive in a few unplanned neighbourhoods; the set of details considered and the quality of end product vary. Usually the need for public land is not considered in advance. The land for public purposes is allocated in the same way as the housing plots. It depends on the timing of public projects (mosque, school, etc.) and if there is land available. Mosques are built in both areas during the early stages of development. In Kibweni, two landowners have left space for the football field. The local team was told to be relatively successful since football is a popular sport in Zanzibar.

None of the land allocators (within the study areas) who were interviewed, stated that they had made a plan of the plots on paper nor had they co-operated with other land allocators. The planning was, however, a reality. An extreme example of local planning was in Mto Upepo Shehia, where one could easily spot six long rows of houses with narrow roads (2-2.5 metres) between them. The seller of the land, together with a local 'mzee' (a respected, usually elderly men), showed the location of the first plots in line by using ropes. All the plots to be sold later were to be in line with these. The allottees were also told how to build on the plots. Similar planning examples were reported by dwellers of the Mtoni Shehia.

All the (informal) allocators of the land admitted that they had some rules they obeyed in the land allocation. Several rules could be listed:

1. copy official size of high density plots;
2. leave minimum space between the houses (3+3 metres);
3. consider the access to houses;
4. avoid selling property in areas likely to flood;
5. optimise the number of plots for sale;
6. sell the plots in order;
7. respect the main internal roads;
8. put the plots in line.

As seen above, not all of the land use decisions that are fixed during the planning phase, are fixed during the land allocation process. After allocation, the private decisions of, and interaction between, the members of new community become crucial. For instance, in the planned area mentioned above, the plan was not visible after one year. Several people have ignored the advice on boundaries and had built on the roads.

The allottees made the land use decisions themselves. The planning decisions they make are often combinations of rational decision making and adaptation to the surrounding solutions. If house drawings are used, the expert has his say in the planning. The planning decisions are co-ordinated in limited number of examples. The most common explanations to private house development decisions were:

1. road (main building and door are usually directed to face main road);
2. neighbour (people either negotiated with neighbours or followed their example);
3. hot climate (many wanted to avoid the hottest sun).

Negotiations with neighbours (Myers, 1993) are valid institutions in designing houses and, especially, when positioning roads and paths. According to the questionnaire, in 20% of cases the landlords defined the place of the road and in 30% of the interviews, the road was there already. 22% reported that they arranged the matter either with neighbours, whilst an individual decision was made in 20% of cases. In Mto Upepo the locations of main doors and pit latrines were negotiated between several houses at the same time to avoid placing the two side by side. Several houses in the row have followed this example. The natural preconditions for the negotiations to be meaningful are the correct timing and attitudes.

Tenure, security and disputes

Tenure in unplanned areas. Land tenure[4] and security of tenure are the main land management elements in unplanned neighbourhoods. In the unplanned areas, the traditions and laws of land tenure are often bypassed and people subdivide land without having right to it. The problem of the order of strength of different claims to land and developments is, however, not a new problem in Zanzibar town (Middleton, 1961; Sanger, 1967; Fair, 1994; etc.). The history of tenure development is partially based on the solutions of conflicts.

Some new tenure conflicts have arisen in the urban periphery. At its extreme, the informal housing markets violate not only the public rights but also individual land rights. Land allocation has become a lucrative business and in some cases the borrowers and keepers of rural land have sold the property even, in the case of Mtoni Shehia, people who have no right to it sell the land of the government. There is an old practice that the cultivators are compensated for the loss of the developments when the owners sell the land, which is leased or borrowed (Fair, 1994; Interviews). Traditionally, in rural towns, the owner of trees has been considered to have the strongest claim to land (Törhönen, 1997). The owner of cultivations does not have similar rights to land.

The allocators have been categorised not in legal terms but in relation to the legal right holders. This indicates that the basis for the present land tenure in informal areas is complicated and further attention to the issue is required before formalising the informal sale of land could take place. The security of tenure of a house builder may be in conflict with the rights of original landholders. When more aspects are taken into consideration, e.g. equality, infrastructure development, building rules, legality etc., the problem becomes much more complicated.

The allocators we met fell in the following categories; they have either:

1. a clear interest to rural land; government lands, three acre plots and private farms;
2. right to the land, which is handed over by the holder of rightful interest. The inherited land is considered to belong to this category;
3. right to use the land, given by the legal holder of land; this includes lands which are borrowed, leased or looked after by someone for the legal holder(s);
4. possession of land to which the right is not handed over by the owner of interest to land nor to trees on it; for instance owners of plants without a right to cultivate or a person who has assumed right to land without having any relation to the grantee;
5. an allotted plot (with or without a building).

Outside the study areas three specific groups of land sub-dividers were also identified:

1. people indicated several cases where the managers of the government land had privately subdivided it for sale;

2. someone subdivided an allocated industrial plot for housing purposes while the allottee was abroad (Saad, 1996);
3. COLE officials (Mohammed, 1996; Ghalib, 1996 etc.) and the sheha in Chuckwani indicated that people have sold planned areas after the declaration of planning area.

The conflict of interest between the state and the allocators of semi-urban land and housing plots around Zanzibar town is among the main problems in unplanned development. This is most evident on government land, where unauthorised users have transformed their cultivations (on a government farm) into housing plots. This is a problem to the allottees that according to the law are unable to apply for the adjudication[5]. According to the local Sheha, the allocators and allottees have been aware that the land and trees belong to the government. Some allottees along the main roads have sought the consent of the Forest Department to cut down trees after they have built their houses.

The law prohibits rural landholders the right to subdivide and sell their properties (LTA, 1992). The policy of not compensating the land in case of expropriation has motivated fast land sales in planning areas (e.g. in Chukwani, 1995)[6] and the conflict has led to demolitions of houses in Mtoni and Mto Upepo during 1996. People had built on the buffer zone of a water-catchment area and the houses had to be removed. A negotiated process between the inhabitants and the state resulted in allocation of plots to several people (Muhajir, 1996) but many were reported to have been left without any compensation after the demolitions and this led to several attempts to squat in neighbouring areas. In June 1997 several houses were again marked for demolition without compensation.

When asked by the people to define which were the important actors or institutions for the development of the study areas, the government offices were not among the most important ones (Group discussions, which took place during 1996). The answers (Table 1) seem to be related to the problems of everyday life, different roles of people (gender) and to local development activities. In Mtoni the definition of the influential actors was too sensitive a question to be handled in detail in a public discussion. Unlike a similar discussion that took place in Kibweni, no private persons were identified by name.

Securing tenure. The Sheha and other local leaders are considered to be the institution for the security of tenure and dispute settlement. More than 60%

supported the Sheha to have the power to settle the local land disputes. The attitudes of the Shehas towards controlling land uses were mixed, although each of them had used their powers in the areas.

The security of tenure is assured in three principle ways on a local level. One may use a witness to certify the seller's right to sell (interview with the Sheha of Bububu in 1996) and to witness the sale of property. For instance, friends, relatives and local leaders witness the transactions.

The other way to secure the tenure is to write an agreement on the sale of property. Some Shehas or their assistants participated in writing and witnessing such papers. Official declarations of ownership of houses and other property were rarely made and 60% were strictly opposed to a statement that they do not need a certificate on their property. Almost the same number of people had some agreement paper.

The third way to secure the right to land is simply to use it and build on it. The method is old and it is considered to work even against the state (Group 5, 1992; Myers, 1993; Andriananjanirana-Ruphin, 1995). The building of a wall and a foundation used to be enough to keep outsiders away (Saad, 1996). Some recent unauthorised sales of developed plots, however, rise doubts on the future of this system. Many walls and foundations are built all over the new informal housing areas of Zanzibar.

Table 1 The three most important people and institutions having an impact on the development of the area.

Area/ group	Women group	Men group
Kibweni	- Midwife - Hospital - Shop	- District officer - Assistant Sheha (local leader)/wife - Representative in Parliament
Mtoni	- Ourselves - No one	- CCM Party Branch - Sheha - Representative in Parliament

Source: Group meetings, 1996

Problems of tenure. The security granted in this informal manner is relative and does not hold against the state as seen above. The sense of security of tenure has usually been considered high in Zanzibar urban areas but after the demolitions, people reported a dramatic decline of land and house prices and of house development in Mto Upepo, with some 33% being afraid of eviction. In Kibweni, which lies 5 km away from the demolition site, only 9% reported the same fears (questionnaire conducted during 1996). Despite this insecurity, people have continued to build in unplanned areas.

The discussions and interviews indicate that there have been only a few boundary disputes in the two study areas. Only 8% of those who answered the questionnaire reported an encroachment of plot, road or open space. Some discussions, however, revealed that such cases were sometimes hidden from strangers. Only 7% demarcate their boundaries. Many plots were sufficient only for houses and a small yard and the demarcation was unnecessary for the owners.

The Sheha of Mtoni reported in 1996 of about ten recent attempts to invade land within his Shehia. In each case the invader came from a house that had been demolished and the invasion was rejected. These examples of securing rural tenure raises a question of whether the invasion of an urban housing plot is any problem in Zanzibar town. No cases of invasion of a housing plot were reported. Two Shehas reported double allocations by one seller of land. This is one of the major tenure problems in these areas.

Some people reported that they (or someone they know) have not challenged their neighbours due to land disputes. Some did not want to disturb the good relationship with the neighbour. Others told that the neighbour is more powerful. Local leaders do not automatically interfere into disputes, unless the matter is found to be very important. Some people are not aware of their rights. There is no study how unbiased the settlements of land disputes by the local leaders and the courts are. Few descriptions of cases, however, reveal that there may be problems.

Other Disputes. Discussions with the Shehas revealed that there are more disagreements on land use than boundary disputes in the two areas. The dispute may concern the blocking of access, the digging of a pit latrine too close to the neighbour's house, the tapping of water without the owner of the pipes permission etc.

The inhabitants and allocators have a limited ability to control the local development. Many pieces of advice and agreements made were violated. The implemented dimensions of buildings were in many cases different from the agreed. In several places, when there was likely to be little requirement for motor vehicles or some other forms of road usage, another form of usage could be permitted (cultivation, installation of structures) providing access was open or could be opened when needed. Among the intolerable problems were blocking the main internal road, digging of pit in a disturbing place, blocking of access and unauthorised tapping of private service connections.

The main problems of the study areas

In the two study areas the most pressing problems are only partially caused by the lack of land use planning and management. Pure land management problems rarely came up in the discussions unless presented by the researchers.

The lack of tap water and malaria were ranked to be the worst problems in both of the study areas. As many as 76% considered the lack of water to be a serious problem (questionnaire conducted during 1996). There were no water taps in the Mto Upepo study area and few in Kibweni, where the taps rarely functioned. The water supply system of the town is not sufficient (Seppänen, 1996a). The finance of the Water Department has been inadequate, partly due to weak national economy. The attempts to introduce water fees to households have also failed. Malaria is a common disease in tropics.

The main problems expressed by the inhabitants concerning the acquisition and development of the land were the financial problems. The group of men from the Mtoni Shehia estimated the minimum plot prices to have been 15,000-230,000 TSh[7] in the early 1990s and the maximum price as 2,000,000 TSh in 1996. It seems that the prices are getting higher. In remote places the value of land is lower. The values of houses were, on average, five times the value of land. The salaries of government experts were about 20,000-30,000 TSh/month[8].

Almost 50% of the interviewees explained that they bought the plot because of the low price or simply because it was the only plot available. In general, the search and purchase of a plot is considered to be a difficult task. Formal plots were considered to be even more difficult to obtain. The problems of affordability would require further investigation.

Men and women have a different ranking for the problems (Table 2). For example the group of women interviewed in Kibweni found the problems related to health care most pressing. They considered the hospital and midwife more relevant to the development of the area than a Sheha or an electrician, which they mentioned as well. The first problem the women would like to solve is the building of a nursery school. The group of men from the same area was more concerned with the poor quality of roads in the area. They felt that there is no institution to help them. They cannot influence project allocation in the road department.

Table 2 The three most pressing problems identified by the dwellers

Area/ group	Women group	Men group
Kibweni neighbourhood	- Lack of money (for medicine) - Lack of medicine - Malaria	- Road problem - Access to electricity - Lack of work/ money
Mtoni Shehia	- Lack of tap water - Lack of drainage - Lack of nursery school/dangerous route to present nursery school	- Lack of water taps - Demolition of CCM - friendly shops - Nursery school is too far

Source: Group meetings 1996

It was easy to observe that in the land allocation process the safety aspects and the need for public land were often ignored. The land use decisions are not co-ordinated and environmental problems are also common. Land buyers don't consider public issues while they look for a building site. Any community and neighbourhood agreements take place after the major land uses are already settled and the Shehas are reluctant to interfere with the details of agreements between the seller and buyer of land. The formal infrastructure experts come to the areas too late for meaningful planning, because they depend on the finance from the inhabitants. When the land for services and infrastructure is needed, it is not available. Security problems are discussed earlier in the text.

Meetings with the experts revealed that they are far from ignorant of the problems in the new, unplanned areas (Seppänen, 1996a; Saleh, 1996; Muhajir, 1996; Khadija, 1996 etc.). In practice, many consider problems in other areas more pressing. Experts also depend on political will. Several institutes are following the priorities set by the leaders of the donor agencies who have little contact with local people.

The attitude towards official land management

The inhabitants of unplanned areas have for years been allowed (or left) to develop the town without official land control. Although the official plot allocation became more effective in the 1990s, people were not aware of many official land management activities. They were not aware of the new land legislation and the rights, which they have. Before and following the elections in 1995 people in some neighbourhoods united their power to impact the official land management (expropriation of land and evictions). The results were both promising and disappointing to the dwellers.

There is only one recent project of the planning without expropriating the land (Muhajir, 1997). The expropriation of rural land for housing without compensation is against the benefits of the rural landholders[9]. The informal land allocation is partly motivated by the fear of loosing the land without compensation. The opinions of the landowners on urban planning are obviously related to this expropriation practice and the expenses of using experts also impact their opinions. No allocator of land interviewed was considered to have any problems in their methods to allocate land. They would continue the allocation in the same manner. Everyone admitted that some planning is needed.

The inhabitants in unplanned areas are in a difficult situation when discussing the need for planning. Housing areas are already built and traditionally up-grading has caused the demolition of some houses. Few expect the government to be able to help them to solve the local problems. 57% of the inhabitants in both areas felt that the lack of land use planning, in general, is a problem, whilst 42% felt that it was a minor problem or not a problem at all. In the same way 57% have a positive attitude towards government planning in the areas while 14% have a strictly negative attitude and the rest do not express any opinion. There was little correlation between the attitudes related to the lack of planning and the attitudes to the introduction of government planning. Many of those who answered gave the impression that they have little to say to official land management. The violations of the local plan shows that the planning *per se* is not respected by all.

When asked what kind of planning people need, the answers were related to infrastructure problems. The water, electricity and road problems were to be solved first. Obviously the planning of a road network alone would help new neighbourhoods. The requests for controlling the locations of buildings came up only twice.

Community development was a reality in both of the areas. Despite this, almost half of the people (48%) fully agreed that those in the area do not plan together. It may reflect the political situation, the uncertain atmosphere due to the bulldozing of houses and undeveloped sense of local community. In Kibweni the opposition was strong; 41% totally disagreed with the claim that people don't plan together and 29% fully agreed. The attitudes of the Shehas to the official planning varied. In general they were quite neutral. Many of them witness sales and write agreement papers. They help people and get some extra income. Most of them see no problems in the present process of land allocation. One Sheha considered that building houses too close to each other was a problem and he was trying to influence the builders. The Shehas usually considered that, in case of official intervention, little should be done for the built houses. One valued rural life and was not enthusiastic about urban development.

Conclusions and recommendations

This research revealed that in new unplanned housing areas there is a set of local procedures to transfer land into housing plots, secure the tenure and solve land disputes on a local level. It also showed that the unplanned development has several problems, which the local community has not been able to solve. The land allocation process and attitudes to land matters differ within one town and within a neighbourhood. Although the informal land allocation system is far from perfect, problems other than land management are felt to be more urgent in the housing areas. In the following paragraphs the informal land allocation process in the study areas is evaluated in relation to some general land policy goals.

The process of the Zanzibar informal plot delivery can compete with the official one in speed. Land officials reported that some people have waited more than 10 years for their plots. Even two years is discouraging for those who need a home. The time required for the acceptance of the building plans also encourage people to buy an unplanned plot.

Official estimations support the idea that the informal process also produces more plots than the formal one. There is, however, lack of information on the total need of plots of different prices and on the supply of them in the town. This issue should be clarified. The interviews and discussions gave the impression that for the lower income families the price of land (not to mention the building) may become a critical issue. Also the fact that many people inhabited dangerous housing units in the Old Stone Town before the evictions in 1996 and that many people build their houses to areas which are not suitable for housing rises doubts that the production of housing plots is not sufficient compared to the need.

Informal land delivery is favourable to the unpredictable income flow of the people. The lack of building control makes it possible to develop step by step and according to the income. This does not mean that all people use the cheapest materials. In formal housing areas the land should be allocated free but the development of officially defined value should take place within two years after the allocation of plot (Haji, 1997). Many are not able to fulfil this requirement.

In unplanned areas, the holders of lands are compensated for their land and developments, which allows them to improve their quality of life to some extent. (They did not seem to be very affluent). The government is not able to tap money from the sales although the highest prices of lands are also related to the proximity to services provided by the government. On the other hand for the majority of the plots it is not providing any services. The provision depends on the customer's finance.

The informal land management does not provide any immediate cost to the public sector. The salary of a Sheha is the only investment made by the government, which to some extent supports the informal land delivery. In the long run unplanned neighbourhoods are expensive. The plot by plot selling of land combined with house by house servicing causes unprofitable investments on the infrastructure. Also building on dangerous sites, waste of best agricultural land and environmental problems is expensive in the long run. Due to lack of local methods to avoid uninhabitable plots and expensive development the development expertise available on the official front should immediately be directed to support the informal process of land delivery. Haji (1997) suggests that it is less expensive to prevent the problems in advance. He considers that the value of land is underestimated, when it is stated that the officials cannot interfere due to the lack of finance. On the other hand the official land management costs could rise the price of land, which would effect the affordability.

The informal land allocation process has not caused the total segregation of people with different incomes to live in different housing areas. In the study areas the maximum household income was 30 times higher than the minimum income. There are a few examples where the services acquired by the wealthier benefit other people as well. In Zanzibar town there are also several high-price, housing zones. Also at neighbourhood level the best plots for sale are affordable only to the most affluent.

It is quite difficult to estimate the effect of unplanned land allocation on the equality of people. People simply perform according to their resources. A close connection with a landowner or good negotiation skills may help to get a proper plot with a low cost or free of charge. The equality related to dispute solution might be a problem.

The official registration of plots usually offers opportunity to use the land as collateral. At the moment an informal plot cannot be used as collateral. Thus existing finance institutions have almost no role in financing house construction in the study areas. The National Housing Bank is bankrupt. It is not clear if new bankers can be attracted to finance low-cost or even middle-cost housing in near future even if the plots would be registered.

The informal land allocation process includes an opportunity to secure the tenure with a relatively low cost and in short time at the local level. The security of tenure is not perfect; it does not protect against the decisions of the government, absentee landholders may lose their land in unauthorised allocation and double allocation is a problem (of limited magnitude). The planned plots are also in danger to be sold by unauthorised sellers. The system of official witnessing of agreement papers should be improved and encouraged, if the land register is not expected to cover these areas in the near future. Even if formal adjudication takes place, there is a need for information on local ownership agreements. Ignoring them is hardly an option. People should immediately be informed about the new and old land legislation in Swahili language to enable them to protect their rights.

People considered the local leaders to be the main settlers of the land disputes. They have solved both land use and boundary disputes. At the local level the power relations of the individuals in dispute may have an impact on the security of tenure. The weakest of society may face problems to get justice. The problem is not limited to unplanned areas. Local leaders or other local representatives should be helped to improve land control methods.

Some land allocators plan their areas before allocating. The quality and quantity of planning/rules depend on the land allocators. In few cases a real comprehensive idea is implemented. The structure of the settlements is fragmented. The public needs safety and the environment is ignored. The role of the individual house owners is strong after the allocation and the lack of development control decreases the impact of planning. There is a need to bring knowledge on planning at the local level.

The results of the study support the idea that Zanzibar's informal land delivery process has some solutions and many challenges to offer for land and development managers.

To be able to utilise the advantages of the informal land delivery process the experts and government officers should create new methods to participate in local land management and for sharing of power and responsibilities. Recent urban land management experiments are promising. The constraints existing within the administration and at the local level should be erased.

To be able to have an impact on development in Zanzibar, the shift of interest from the reorganisation of land management, within the government bodies, to working with the actual support actions should take place. The transition of administration and land use planning laws could be a change to enter new land actors (development experts) to the local level. The discussion on the future land management should be public.

Acknowledgments

The field study has been financed by the Academy of Finland and the Nordic Institute of African Studies. The project "The Research of Land Management in Developing Countries" administrated and directed by the Prof. Kari I. Leväinen in the Helsinki University of Technology has given valuable support for conducting the work.

During my stay in Zanzibar, I have received hospitality and great help from many Zanzibaris, Tanzanians and Finns. Of the many who helped me in COLE and Ardhi Institute I would like to thank Mr Mohammed Adam, Mr Mohammed Haji, Mr Muhajir, Mr Ghalib, P.S. Mohammad Salim Sulaiman, Dr Kombe and Mr Burra for the important ideas and valuable help with work arrangements. During my stay in Zanzibar the support from Mr Veikko Korhonen and Mrs Anja Korhonen was exceptionally warm-hearted and substantial. I also own deep gratitude to the inhabitants and local leaders from Kibweni, Mtoni, Mto Upepo and other visited areas. Their co-operation and kind help was truly motivating. The most crucial, if possible, for the success of the field work has been the intelligence, humour and friendliness which Ms Khadija Masrur Saad from COLE contributed day after day. Asante sana! Asanteni sana!

References

Andriananjanirana-Ruphin Solange, 1995. Zanzibar Town Planning Problems 1890-1939, *The History and Conservation of Zanzibar Stone Town*, Eastern African Studies, Ohio University Press 1995, London, 100-108.

Angel, S., Archer, R.W., Tanphiphat, S., & Wegwlin, E.A. (eds), 1983. *Land for housing the poor*, Select Books, Bangkok.

Burra, M., 1996. *Autonomous Self-Initiated Planning and Local Spatial Management for the Informal Settlements - A Case Study of Makongo Settlement*. An unpublished preliminary report, Ardhi Institute, Dar es Salaam.

Carroll, A., 1980. *Private Subdivisions and the Market for Residential Lots*, World Bank Staff Working Paper No. 435, October 1980.

COLE Act, 1989. *The Commission for Lands and Environment Act 1989*, Act No 6 of 1989, Zanzibar.

Fair L., 1994. *Pastimes and Politics: a social history of Zanzibar's Ng'ambo community 1890-1950, Vol 1*. Doctoral Thesis, University of Minnesota, UMI, USA.

Falleth, E.I., 1993. *Land Actors and Settlement Development; A Case Study in Nepal Kathmandu and Patan, Nepal*, Norwegian Institute for Urban and Regional Research, Note 1993, 117, Oslo 1993.

Group5, 1992. *Identification of Strategy for the Urban Development of Zanzibar Town*, Draft Report, Group5 consulting engineers, Government of Zanzibar, UNCHS.

Haji, M., 1997. Head of Integrated Planning Unit, COLE, Discussion 12.6, Zanzibar.

Lischi, M., 1993. *The Mtoni Water Source - A Preliminary Study on the Threats to an Important Natural Resource for Zanzibar Town*, Zanzibar.

LTA, 1992. *The Land Tenure Act of 1992*, Zanzibar.

Khadija, M.S., 1996. Land Use Planning Officer. Several discussions 15.3.1996-5.5.1996, Integrated Planning Unit, COLE.

Kironde, L., 1995. Access to land by the urban poor in Tanzania: some findings from Dar es Salaam, *Environment and Urbanization, 7*, 77-95.

Kombe, W.J., 1992. *Lessons of Experience from Informal Housing Settlements in Chang'ombe and Chamwino, Tanzania*, UNDP/ILO INT/89/021, Dar es Salaam 1992.

Kombe, W.J., 1996. *Formal and Informal Land Management in Tanzania- The Case of Dar es Salaam City,* Spring Research Series No. 13, Spring Centre, Dortmund.

Larbi, 1995. *The Urban Land Development Process and Urban Land Policies in Ghana,* University of Reading, RICS, London.

Mabogunje, A.L., 1990. Urban Planning and the post Colonial State in Africa: A Research Overview, *African Studies Review,* **33**(2), The African Studies Association.

Merriam-Webster, 1997. *Merriam-Webster's Collegiate Dictionary,* Tenth Edition, Encyclopaedia Britannica, Inc. and Merriam-Webster, Inc. on-line, *http://www.eb.com.*

Middleton, J., 1961. *Land tenure in Zanzibar,* Her Majesty's Stationery Office, London.

Muhajir Makame, 1993. *Settlement Planning Improvement in Zanzibar-Towards a framework for Implementation,* Unpublished Master's Thesis, Curtin University of Technology, Australia.

Muhajir Makame, 1996. Director of Department of Surveys and Urban Planning, COLE, Zanzibar, Interviews 22.3.1996 and discussions in June 1997.

Muhajir Makame, 1997. Director of Department of Surveys and Urban Planning, COLE, Zanzibar, discussions during June 1997.

Muhammed Adam Muhammed, 1996. Head of the Integrated Planning Unit. COLE, Zanzibar, several discussions 15.3.1996-5.5.1996.

Myers, G. A., 1993. *Reconstructing Ng'ambo. Town Planning and development of The Other side of Zanzibar,* Doctoral Thesis, University of California, Los Angeles.

Myers, G. A., 1994. Eurocentrism and African Urbanization: The Case of Zanzibar's Other Side, *Antipode,* **26**(3), Blackwell Publisher, Oxford, UK.

NLUP_S.01, 1995. *National Land Use Plan, Planning Policies and Proposals,* Revolutionary Government of Zanzibar, Zanzibar.

Ochieng'-Akatch, S., 1994. *Evaluative Review of Urban Planning Practice and Experiences in the African Region,* Draft, University of Nairobi, Kenya 1994.

Payne, G., 1989. *Informal Housing and Land Subdivisions in Third World Cities, an overview of the literature,* Overseas Development Administration, London.

PD, 1964. *The Confiscation of Immovable Property Decree,* Presidential Decree, No. 8 of 1964, Zanzibar.

PD, 1965. *A Decree to Vest all Land in Government,* Presidential Decree, No. 13 of 1965, Zanzibar.

PD, 1966. *The Land Distribution Decree,* Presidential Decree, No. 5 of 1996, Zanzibar.

PD, 1969. *A Decree to Amend the Land (Distribution) Decree, 1966,* Presidental Decree, No. 1 of 1969, Zanzibar.

Saleh, 1996. Executive Water Engineer, The Department of Water Development. Interview 20.3.1996, Zanzibar.

Sanger, C., 1967. Introduction. In: Okello, J., *Revolution in Zanzibar*, 6-22, The African Publishing House, Nairobi.

Singer, N., 1991. *Report on Land Use Planning Soil and Water Conservation and Land Tenure in Zanzibar*, Consultancy Report, Alabama, USA.

Singer, S., 1993. Part I. In: Singer & Törhönen, *The Adjudication Planning Process in Zanzibar; Investigation of Seven Land Tenure Types*, ZILEM, Zanzibar.

Seppänen, O., 1996. *Zanzibar Town's Population Projection for Years 1990-2015*, Unpublished paper, Urban Water Supply Project 1996, Zanzibar.

Seppänen, O., 1996a. Chief Technical Advisor, Water Department. Urban Water Supply Project, Interview 20.3.1996.

Symposium on Shelter, 1991. *Symposium on Shelter in Sustainable Urban Development in Africa Recommendations to African Governments*, 15th May 1991, Gaborobe, Botswana.

Törhönen, M., 1997. *Land Tenure Confused: Past, Present and Future of Land Management in Zanzibar*, Unpublished dissertation 1997, Helsinki University of Technology.

Yahya & Associates, 1982. *Land Policy for Zanzibar and Pemba*, Final Report. Ministry for Land Construction and Housing, Zanzibar.

Yahya & Nzioki, 1994. Unpublished, Kenya 1994.

ZMCA, 1995. *Zanzibar Municipal Council Act 1995*, Act #3 of 1995, Zanzibar.

Endnotes

[1] Zanzibar town is the main administrative and service centre of Zanzibar. The town is estimated to have 193,700 inhabitants while about 40.000 urban or semi-urban dwellers live outside the town boundaries. The growth rate of the population in areas outside the town is estimated to decline from 11% today to 7% by 2015. By 2015 the urban areas around the town would have as many inhabitants as the town today (Seppänen, 1996).

[2] Expropriated (after declaration of planning area) land has not been compensated.

[3] People have freely united their powers with neighbours to gain services. Usually the local leaders are organising neighbourhood level organisations either through the Sheha institution (local leaders acting between the people and the government administration) or through the party institutions (old projects). The installation of basic services (water, electricity) has united people in the neighbourhoods of the town. The land tenure is not usually impacted in these projects.

[4] All land in Zanzibar is public, owned by the state and at the disposition of the President (PD 13/1965, LTA 1992). The state has had the monopoly to control the allocation of urban plots since the revolution (PD 1/1969, see also COLE Act 1989, Zanzibar Municipal Council Act 1995). Land tenure is not uniform. One can separate three types of officially

defined land tenure types around Zanzibar town relevant to housing development; rural three acre plots (holdership for lifetime, conditional, inheritance trough application) which were allocated after the revolution (PD 5/1966, LTA 1992), agricultural properties which have remained in the hands of the pre-revolutionary owners and the state properties (PD 8/1964, see more in Yahya, 1982a; Singer, 1991; LTA, 1992). The new land legislation introduced a Right of Occupancy to land, which is planned to unite the tenure. The same law states that the rights to land and to developments can be held separately (LTA, 1992).

[5] According to the Land Adjudication and Land Registration Acts a person is entitled to apply for the Right of Occupancy after 12 years of peaceful and undisputed occupancy of the land. The law strictly prohibits the right from those who have settled to land, which is registered to the government.

[6] The demolition process was long; the water problem was realised years before the evictions took place (Lichi, 1993). Some people criticise the government for not controlling the building early enough. There are also political speculations involved.

[7] 1 USD = 615 Tsh (1997)

[8] The salaries are inadequate to cover the living expenses.

[9] New Land Tenure Act introduces due compensation for the land.

The utility of some synoptic information tools for the environmental sustainability of land management strategies in the Asian context

Dr. Amador A. REMIGIO
School of Environment, Resources and Development
Asian Institute of Technology, Thailand

Abstract

During the last decade land management strategies have been developed to ensure a policy of environmental sustainability. The problem of developing such strategies in the Asian context is that the regions has been described as data and information poor and thus the challenge has been to ensure that the collection and analysis of data is timely, relevent and reliable. Furthermore, it must be ensured that both the quality and quantity of collected data and processed informationis adequate for land management purposes. In order to ensure that this data is used in the most effective manner it is, therefore, required that stakeholder-related and environmental sustainability information be used to determine the shaping, as well as implementing policies and strategies for the management of natural resources such as land.

This paper focuses on three inter-related facets of complementary land management policies: the role of the stakeholder in the development of environmental sustainability; the ability of land management strategies to incorporate environmental sustainable considerations; and the determination of whether some synoptic information system tools may be used to generate learning experiences to stakeholder-sensitive, environmentally sustainable, land management strategies.

Introduction

In a recent strategy publication relating to the development-environment interface, environmental sustainability strategy was defined as

> a process of designing and taking a set of actions to strengthen or change values, knowledge, technologies and institutions to achieve specific objectives with the ultimate goal of improving and maintaining the well - being of people and ecosystems (Carew-Reid et al 1994).

Simply put, such strategies are

> processes of planning and action to improve and maintain the well - being of people and ecosystems (Carew-Reid *op cit*).

From a recent survey covering 14 years of experience with the formulation and implementation of such strategies, the following lessons can be gleaned (Carew-Reid *op cit*):

- these strategies seek to improve and maintain the well-being of people and ecosystems
- the overall goal of these strategies is sustainable development
- the choice of strategy objectives should be tactical
- the strategy process is adaptive and cyclical
- the strategy should be as participatory as possible
- communication is the lifeblood of a strategy
- strategies are processes of planning and action
- strategies should be integrated into the decision-making systems of society
- the capacity to undertake a strategy must be built at the earliest stage
- while external agencies have a legitimate role to play in such strategy formulation and implementation, they should be in a position of being readily 'tapped' for the strategy, rather than being on top of the strategy.

The 'greening' of the 'consciousness' of development and environment professionals came to be reflected in the belated recognition of two particular considerations. Both wise planning and decision-making, relative to the formulation and implementation of these environmental sustainability strategies, is contingent on the availability of systematically organised information on both the biogeophysical as well as socio-economic aspects of the environment and natural resource base. (These information sets have been made available primarily through research, participatory processes and the interchange of information through communication). Thus, there is a compelling need to synthesise both biogeophysical and socio-economic data and information for environmental sustainability strategy formulation and implementation in an integrated, synoptic form. Such data and information form or "structure" can then serve as useful decisional inputs that can help both the planners and managers of various development activities, projects, plans, programs and policies that have resource and environmental implications. Likewise, the multiple stakeholders affected by the implementation of all these anthropogenic, 'developmental' interventions will have to be influenced towards the realisation of environmentally sound outcomes.

The situation in Asian developing countries has been instructive in this respect. To the extent that these countries have been characterised as 'data and information poor', the challenge therefore has been not only confined to ensuring that such collected and analysed data and information is timely, relevant and reliable but to ensure also that both the quality and quantity of collected data and processed information is adequate for natural resources and environmental planning and management purposes (e.g., land management). Because the improvement and maintenance of the welfare and well-being of people and ecosystems is the *raison d'etre* for undertaking sustainable development, it is reasonable to presume that natural resource and environmental (including land) management strategies should also incorporate both major stakeholder orientations and environmental sustainability considerations for addressing priority natural resources and environment-related issues. Such deliberate factoring of both crucial stakeholder concerns and environmental sustainability aspects cannot be underestimated in terms of their significance in re-orientating natural resource and environmental management strategies towards more viable outcomes in terms of upholding the welfare and integrity of peoples and ecosystems. What these requires therefore is that relevant stakeholder-related and environmental sustainability information be used in determining and shaping the crafting as well as execution of policies and strategies for the management of natural resources such as land.

The focus of this paper will, therefore, be on:

1. the role of stakeholder orientation as a crucial variable between knowledge and action in incorporating environmental sustainability considerations into land management strategies;
2. the receptivity of land management strategies to incorporating environmental sustainability considerations from scientific and technical knowledge inputs in conjunction with the role of information system tools in enriching such knowledge inputs; and
3. on whether some synoptic information system tools can generate learning experiences valuable to stakeholder-sensitive and environmentally sustainable land management strategies.

The role of stakeholder orientation as a crucial variable between knowledge and action in incorporating environmental sustainability considerations into land management strategies

For the purpose of this paper, the IUCN, UNEP and WWF definition of what constitutes environmental sustainability in natural resources and environmental management (1991) will be adopted. As defined, natural resources and environmental management in an environmental sustainability mode calls for

> the protection of natural ecosystems, the production of wild renewable resources from modified natural ecosystems, the sustainable production of crops and livestock from cultivated ecosystems, the development of built (or urban) ecosystems in ways that are sensitive to human and biological communities and the restoration of degraded ecosystems.

Thus, natural resources and environmental management that takes on board environmental sustainability considerations has been characterised as essentially adaptive

> a way of managing in order to ensure that the organisations responsible for ecosystems are responsive to the variations, rhythms, and cycles of change natural in that system and are able to react quickly with appropriate management techniques (Westley 1995).

Its adaptive nature is particularly allied with a peculiar image of the relationship of the organisation to its environment, an image which suggests the organisation being perceived as a machine that had rational, analytic and programmatic strategic planning as its engine (Bowman 1990). Notwithstanding these, there has been a discernible paradigm shift to an organisational image that finds organisations as 'embedded' in their environments (Granovettor 1985), with strategising being viewed as a more naturally emergent process that results in meanings (and their interlinked learnings) within the system. This change in focus from control to responsiveness that such paradigm shift entails has meant the reconsideration of the function of planning as well as a renewed pursuit of alternative options for generating such meanings and learnings (Westley *op cit*) relative to a sustainable development context.

A caveat though must be raised at this juncture. It has been observed that the frenzied attempts to improve the availability and analysis of information about biogeophysical and socio-economic environmental resources has been anchored on the assumption derived from classical economic theory that the rationality driving such attempts will necessarily result in informed (and, therefore, better) decision-making. Notwithstanding this assumption, an increase in the quantity of information available and the provision of qualitatively better information into natural resources and environmental management decision-making, in an environmental sustainability mode, does not necessarily lead to better decisions, plans and management practices in the natural resources and environmental planning and decision-making arena (Hassan & Hutchinson 1992).

It must be qualified that information could conceivably be just one among a constellation of factors affecting such natural resources and environmental planning and decision-making. Such factors can include explicit objectives as well as implicit motives, intangible biases, preferences and inclinations, of the planners and decision-makers, political considerations and cultural imperatives. Moreover, planners and decision-makers often face formidable constraints such as ill-defined problems, unclear objectives, pressures from various quarters, information inadequacies, the non-systematic consideration of all the possible options and the consequent atmosphere of risk and uncertainty surrounding such problem-solving and decision-making. Coupled with these developments have been the tightening of environmental regulations by international, national, regional and local agencies. In addition, there is an increasing necessity to undertake the evaluation of the interactions and linkages in the development-environment interface within and across areas and regions, twin developments that require an increasing capacity to integrate and synthesise data and information from varied sources.

An illustrative example that can be used in this respect is the recent phenomenon of forest fires (during the last two quarters of 1997) in the Indonesian islands of Kalimantan and Sumatra. While aerial reconnaissance photography and satellite remote sensing afforded a geo-synoptic view of the extent of the haze that these forest fires have produced, the management response of Indonesian decision-makers have had to reckon with a plurality of major factors that have conditioned such response.

Among these are:

- the motives of the stakeholders (in terms of logging and plantation companies and farmers explicitly resorting to the use of forest fires to clear land for agriculture and agro-forestry purposes);
- the government's own biases, preferences and inclinations (reflected in the failure to acknowledge that the Indonesian government was partly responsible for the forest fires because of its lax regulation of these shifting cultivators and logging and plantation companies while alternately pinning the onus of guilt on both these shifting cultivators as well as logging and plantation companies);
- political considerations (as manifested in the rather belated government response to address the problem only after the outcry from neighbouring countries over the haze became transnational and regional in character);
- cultural imperatives (with the government later issuing an apology to the affected countries but nonetheless continuing to duck responsibility).

The receptivity of land management strategies to incorporating environmental sustainability considerations from scientific and technical knowledge inputs in conjunction with the role of information tools in enriching such knowledge inputs

To what extent then can the formal natural resource and environmental planning process are receptive to the consideration and incorporation of critical and relevant scientific and technical knowledge inputs? While the incorporation of such knowledge inputs into planning can be viewed as being contributory to laying the groundwork for subsequent action, the receptivity of such formal planning process to such inputs will be contingent on:

- the form and usability of such knowledge for natural resources and environmental planning and decision-making, and
- the relative strength and dominance of the organisational paradigm that is directing both the planning (and decision-making) process (Westley *op cit*).

The provision of geo-synoptic information and knowledge in cartographic form to decision-makers is definitely helpful in spatially defining and delineating the intensity, scope and magnitude of the impacts of natural resource depletion and environmental degradation. This may be considered as a preliminary step in natural resources planning and decision-making that consciously incorporates environmental sustainability considerations:

- in order for environmentally sustainable natural resources and environmental planning and management to be effectively undertaken, these should be based on viable scientific and technical knowledge inputs (Bakkes *et al* 1994), with these inputs being systemically related to each other (as illustrated in Figure 2);
- the current status and distribution of biogeophysical and socio-economic resources (including trends over space and/or time);
- the pressures being exerted on these biogeophysical and socio-economic resources ;
- the effectiveness of current policies and management responses.

Thus, information regarding these inputs would be definitely required in order to effectively undertake natural resources, environmental planning and decision-making activities on an environmentally sustainable basis. It is interesting to note that all these information and knowledge inputs address the questions raised by Hammond *et al* (1995) as pertinent to the analysis and measurement of environmental policy performance in sustainable development, such as "what is happening?", "why is it happening?" and "what are we doing about it?". As a heuristic example, in the context of crafting stakeholder-sensitive and environmentally sustainable land management initiatives, stakeholder analysis could possibly indicate that specific community segments (e.g., the rural poor) are relatively disadvantaged in their access and control to specific forest resources that are vital to their livelihoods. The indiscriminate cutting of trees (with the rate of harvesting and exploitation being much greater than the rate of biomass growth and replenishment) leads to ecologically unsustainable resource use patterns. As a consequence, forest resource depletion and degradation could very well be linked to stakeholder disparity in access and control to such natural (forest and land) resources. Thus, subsequent policy and management responses that are not informed by adequate scientific, technical and social knowledge inputs could result in outcomes that treat the symptoms rather than the underlying causes of such forest resource depletion and degradation.

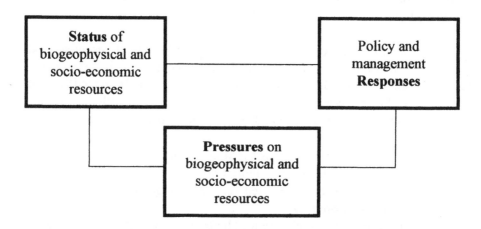

Figure 1   **Scientific and technical knowledge as inputs to environmentally sustainable natural resources and environmental planning, as modified from UNEP and WCMC (1996)**

There is this growing realisation then that such scientific, technical and social knowledge inputs increases the level of understanding. Similarly, analysis clarifies the nature as well as dynamics of the environmental issue and enables the planner and decision-maker, as well as other stakeholders, to be empowered to make socially optimal and environmentally sound decisions. It must be noted though that the eventual impact of such knowledge inputs on such planning and decision-making will be a function of the way in which the information is presented. This will determine the extent to which it is immediately relevant to natural resource and environmental planning and decision-making needs (UNEP & WCMC 1996). Such realisation therefore brings to the fore the issue of the wisdom of making investments in ensuring that these knowledge inputs are systematically collected, stored, retrieved, analysed and probed for their practical significance to stakeholder-sensitive planning in environmentally sustainable natural (forest and land) resource strategy formulation and implementation.

Such investments must be justified in terms of their cost-effectiveness in serving natural resource and environmental planning and management needs as well as resolving crucial issues at the policy, program and project levels. The servicing of such needs and the resolution of such issues will entail, therefore, the involvement and participation of a broad spectrum of stakeholders. These will include (but not be limited to) national and local government officials, natural resource and environmental management agencies, NGOs, local people and community groups, the private sector, the academe and other significant segments of the polity, economy and society. In the case of the interface between natural resources/environmental planning/management and these multiple stakeholders, the integration and synthesis of varied perspectives and viewpoints with numerous data and information sources necessitates the use of efficient and effective information management tools.

However, the peculiar structuring and compartmentalising of data and information sets into biogeophysical and socio-economic 'boxes' have not helped in facilitating the integration and mainstreaming of both biogeophysical and socio-economic considerations into environmentally sustainable natural resources and environmental planning and management. Moreover, data and information sets from both the biogeophysical and socio-economic fields may be transformed into usable knowledge inputs for stakeholder-sensitive and environmentally sustainable natural resources and environmental planning and management. However, there should be a conscious effort, on the part of both the data and information producers and knowledge users, to be clear about the rationale, significance and relevance of an integrated approach to orchestrating data and information collection and analysis activities and products. In this regard, the information and knowledge products that will then be used as knowledge inputs into stakeholder-sensitive and environmentally sustainable natural resource and environmental planning and management should meet the criteria of:

- being easily understandable (or attuned to the 'wavelength' of its intended audience);
- factual;
- logical;
- problem-focused;
- capable of generating critical and analytic inputs that can helpfully narrow or broaden the range of options for planning and decision-making; and
- contributory to the monitoring and assessment of the differential impacts of planning and management interventions (insofar as these can be generative of feedback information useful for project, program and policy planning and management).

The various synoptic information tools and methods that can be used to enrich these knowledge inputs into stakeholder-sensitive and environmentally sustainable natural resource and environmental planning generally fall into the following categories:

- generic tools for determining information needs (e.g., questionnaires, structured interviews, workshops and working groups, brainstorming, information networking);
- generic tools for facilitating data and information interchange (e.g., communication and collaboration, access to library and research materials in published form and the tapping of online environment - related information sources in the Internet, as accessed by powerful meta-search and search engines designed for the World Wide Web);
- biogeophysical and socio-economic resource information generation, analysis and management tools, such as:
  - survey tools (e.g., informal consensus, Delphi);

- ground survey techniques (e.g., rapid rural appraisal using qualitative individual and group interviews, direct observations, informal structured surveys, systematic ground sampling and area sampling frames);
- geographic-based mapping and cartographic analysis tools (e.g., computer-aided mapping, remote sensing and geographic information systems (GIS), new analytical approaches such as environmental characterisation and modelling by spatial statistics, evolving database management and software designs, e.g., hybrid artificial intelligence systems used as knowledge "discovery" tools that use computers and software algorithms to extract implicit and potentially useful information from data as well as the development of visualisation software) (Michener *et al* 1994);

- textual, tabular and numerical information processing (e.g., word processing, spreadsheet, relational database management systems) that utilise 'cutting-edge' computer hardware and software-based information technologies.

Examples of data and information situations calling for such technologies are:

- data that contain relationships that are too complex to be manually tracked or are too great in volume for conventional filing systems to cope with;
- the imperative of integrating and synthesising data from multiple sources into a combined output;
- the need for data and information to be shared between and among various users within an institution or outside of these institutions;
- the need to periodically or continually search, sort or update data;
- where there is a need to regularly generate information outputs from these databases.

For this paper's purpose, the focus of discussion will be on geographic-based mapping and cartographic analysis tools such as remote sensing and GIS, defined by Estes *et al* (1983) as computer-based systems for integrating and analysing spatial information in a decision-making context.

The more significant benefits associated with the use of these tools in Asian developing countries can hardly be disputed in terms of their role in facilitating development planning, infrastructure project design and natural resource and environmental planning and management. This is particularly so in helping determine the country's natural resource potential through resource inventories that show the extent of such natural resources and their spatial distribution. In the case of remote sensing, its major natural resource and environmental applications in the Asia-Pacific context have been on the following fields:

- monitoring of regional land use patterns and changes;
- monitoring of the distribution of geological structures and soil types;
- erosion susceptibility assessment using data on slope, soil characteristics, vegetation cover and drainage patterns;
- monitoring of short- and long-term changes in vegetation (e.g., density of green vegetation) and vegetation classes and the influence of natural/anthropogenic factors such as pest infestations, forest fires, logging operations, shifting cultivation in causing deforestation for forest (and watershed) planning and management;
- mapping of forest types and the extent of deforestation and reforestation;
- monitoring of quality and quantity changes in surface water sources and flood risk for water resources assessment/ development, flood/drought forecasting and flood-related infrastructure planning;
- the usage of multi-temporal remote sensing data for determining the extent and location of cultivated land, crop forecasting (including area and yield estimation of specific crops) and the tracking and prediction of plant parasite infestation;
- land suitability assessment;
- land (and forest) rehabilitation planning.

In addition, the role of these synoptic information tools has been definitely enhanced by the built-in capability in remote sensing and GIS systems to use a number of analytical techniques for spatial analysis. This may be defined as such insofar as it answers questions relating to geographic feature location, its attributes and the linkages of such geographic features and its attributes to other geographic elements. This includes such elements as, multi-thematic data, information integration and synthesis from various sources (such as satellite remote sensing and aerial images, topographic maps, socio-economic data and other non-spatial data attributes) and the presentation of the analytic results therefrom in various map types.

In the case of GIS systems, their capability in handling and analysing:

> large volumes of spatial and non-spatial data with greater accuracy in less time

allows the visualisation of scenarios and the simulation of the range of potential effects of a 'developmental' or 'environmental' intervention (UN ESCAP 1994). This might typically be on a natural (forest or land) resource and environmental base.

The 'appropriate application' of remote sensing and GIS in Asian developing countries has afforded the opportunity of collecting and processing

> systematic, accurate, synoptic, holistic, repeated and unbiased information about natural resources and the environment (UN ESCAP *op cit*)

in a fairly rapid, efficient and economical manner. The significance of these for natural resources and environmental planning and management cannot be ignored as past experiences have shown that biased, incorrect and inadequate information oftentimes lead to unrealistic strategies (UN ESCAP *op cit*). Notwithstanding these benefits, there have been substantive problems that have served as obstacles that have precluded the full realisation of these benefits (Pheng 1996; Valenzuela 1996; Tewari 1996), such as:

- acquisition and operations cost: A number of organisations in the Asia-Pacific region find it beyond their financial means to acquire commercial remote sensing and GIS systems (in terms of hardware and software components) while those who already have a base remote sensing and GIS system find that the required operational costs (e.g., to be incurred for database development and training) are rather prohibitive;

- data collection and management problems: Many Asia-Pacific developing countries can be characterised as "data-poor" relative to the data sets required by remote sensing and GIS systems. Geographic data inputs for remote sensing and GIS applications are oftentimes:

    - unavailable in terms of data users having the right data at the right time given the "multitude of data product types", as well as numerous hardware and software systems, that make it hard for Asian developing country users to derive the maximum benefit from whatever available data there is (UN ESCAP *op cit*);

    - largely inaccessible because of data incompatibilities in terms of varying formats, standards and calibrations, the trend towards distributed databases and the multiplicity of data sources that are making it

        difficult for Asian developing nations to acquire mission-specific hardware and software to collect, process and analyse all the data being generated by various remote sensing satellites (UN ESCAP *op cit*);

    - unsuitable (in terms of accuracy, quality, reliability and currency).

- institutional constraints, which take the form of data security and confidentiality stipulations reflected in restrictions on the use of large-scale mapped information and aerial photographs that are imposed by defence and national security agencies, inadequate institutional capacity to meet the technical demands of remote sensing and GIS systems (often manifesting itself in the uncritical application of remote sensing and GIS methodological frameworks without adapting these to local site-specific conditions and the lack of knowledge on how to interface such remote sensing and GIS technologies to the overall planning process) and the problem of co-ordinating

> multiple use and multi-agency GIS applications because of these agencies' narrow sectoral mandates;
>
> - the failure to realise that remote sensing and GIS tools also have their own current limitations relative to natural resources and environmental planning and management (e.g., significant delays entailed in the collection, processing, evaluation and interpretation of data relative to natural resource and environmental planning and decision-making applications that require "real-time" information to cope with the management imperatives dictated by active and dynamic environmental phenomena such as the large-scale forest fires in the Indonesian islands of Kalimantan and Sumatra).

Synoptic information tools and sustainable land management strategies

Given this wide array of information tools and methods that can yield both biogeophysical and socio-economic data and information, it is important to understand the information process that underpins stakeholder-sensitive and environmentally sustainable natural (forest and land) resource and environmental management strategy planning. Such process begins with an agreement by the concerned stakeholders and information end-users as to the pre-eminent biogeophysical and socio-economic issues that deserve priority attention. Corollary to these would be the ascertaining of what the information needs of these issues are and in what useful form the information products should be. Then, there should also be a fundamental 'meeting of the minds' in terms of what the stakeholders' roles, capacities and responsibilities should be in ensuring that such information needs are serviced, with the right information products being delivered at the appropriate time.

Upon reaching agreement on what are the pre-eminent issues to be addressed, the assertion of what the information needs are necessarily follows. A basic checklist comprising the following questions (as modified from UNEP & WCMC *op cit*) can then be used as a means to focus attention on what these information essentials are:

- what specific major, urgent and significant stakeholder-related issues in environmentally sustainable land management strategy planning are being addressed?
- which groups in the society, polity and economy exert the major and decisive influences that affect stakeholder relations and dynamics in such environmentally sustainable land management strategy planning?
- how do these groups collect, analyse, monitor and use stakeholder-related information inputs in land management strategy planning;
- what stakeholder-related policies exist relative to such land management strategy planning (or is there a policy vacuum that impoverishes such strategy planning)?
- what specific information inputs are then required to assist in the implementation, evaluation and development of such stakeholder-related policies for land management?
- how, when and to whom should these specific information inputs be assembled, produced and eventually delivered?

Parallel to such information needs determination must be a consideration of what constitutes an 'ideal' information product. It should have characteristics such as:

- being designed for a specific audience;
- relevant to land management planning and decision-making needs;
- available when the "window of opportunity" for land management planning and decision-making opens;
- easily and quickly understood;
- based on sound scientific principles;
- delivered through recognised channels;
- clearly identifies areas of uncertainty and their significance;
- accompanied by full acknowledgement of data sources and intellectual property;
- available at minimal cost in terms of time, money and administrative overheads.

It should be clear then that the design, production and delivery of such an ideal information product, in the process of undertaking stakeholder analysis and enriching the stakeholder dimensions of land management strategy planning, can be the object of various technological interventions. It should be assumed that the general definition of technology as processes by which human beings create tools and methods to increase their control and understanding of both the biogeophysical and the socio-economic environment.

It is in this vein then that the issue of the pertinence of information technology tools and methods, to making natural (forest and land) resource and environmental planning more stakeholder-conscious and environment-ally sustainable, should be examined. Can such synoptic information tools and methods really generate learning experiences valuable to land management strategy planning and implementation as to transform it to be more stakeholder-sensitive and environmentally sustainable?

This paper has attempted to survey and assess the current utility of some synoptic information system tools and methods, specifically remote sensing and geographic information systems, to the implementation of land management strategies that are anchored on a stakeholder-sensitive approach and its pursuit on an environmentally sustainable basis. Given such tools and methods that have been made possible by rapid advances in computer and information sciences and technologies, it can be readily seen that these tools and methods have practical significance in addressing natural (forest and land) resource and environmental issues, problems and concerns. These concepts are central since they are the core to the implementation of land management strategies. However, there are formidable obstacles and barriers that currently preclude the utility of these tools from being fully optimised.

Whether such learning experiences result in positive or negative outcomes will surely hinge on moral choices by the end-users of these information tools. For technology, generally defined, is but a glorified means (made possible by science) specifically harnessed towards the realisation of specific ends that either heighten man's greatness or accentuates his fall. Just as space-based information technologies have been used for peaceful and/or military purposes, information technologies for social engineering *cum* ecosystems protection, conservation and renewal schemes. These would include such as environmentally sustainable land management strategy planning and implementation. Thus they could be used to improve and/or maintain the welfare and well-being of peoples and ecosystems. Alternatively, and more ominously, they could be used as a means by authoritarian governments, for example, to further threaten the welfare, well - being and quality of life of peoples and communities. The values, knowledge and culture may, however, run counter to what the government propounds as well as their life-support ecosystems. There is no escaping then the fact of technology being Janus-faced but the moral imperative remains. It is incumbent upon the designer and end-user of technologies to choose whether the technological means available at his disposal will be used to pursue that which is good, noble and upright. This should be compared in relation to worthwhile sustainable development objectives or to engage in activities that ultimately lead to the overall unsustainability of peoples and ecosystems.

References

Bakkes, J.A., van den Born, G., Helder, J., Swart, R., Hope, C. & Parker, J., 1994. *An Overview of Environmental Indicators: State of the Art and Perspectives - an environmental assessment technical report*, UNEP/RIVM.

Bowman, E.H., 1990. Strategy changes: possible worlds and actual minds. In: J.W. Frederickson, (ed.), *Perspectives on Strategic Management*, 9-37, Harper and Row, New York.

Carew-Reid,.J., Prescott, A.R., Bass, S. & Dalal, C.B., 1994. *Strategies for National Sustainable Development A Handbook for their Planning and Implementation*, International Institute for Environment and Development and the International Union for the Conservation of Nature.

Estes, J.E. *et al.*, 1983, *Advanced Data Acquisition and Analysis Technologies for Sustainable development*, MAB Digest 12, UNESCO.

Granovettor, M., 1985. Economic action and the social structure: the problem of embeddedness, *American Journal of Sociology*, 91(3), 481-510.

Hammond, A., Adriaanse, A., Rodenburg, E., Bryant, D. & Woodward, R., 1995. Environmental indicators: a systematic approach to measuring and reporting on environmental policy performance. In: *Context of Sustainable Development*. Washington, D.C., World Resources Institute.

Hassan, H. & Hutchinson, C., (eds.) 1992. *Natural Resource and Environmental Information for Decisionmaking*, Washington, D.C., World Bank.

IUCN, UNEP & WWF, 1991. *Caring for the Earth: A Strategy for Sustainable Living*. London, Earthscan.

Michener,W.K., Brunt, J.W. & Stafford, S.G. (eds.) 1994. *Environmental Information Management and Analysis: Ecosystem to Global Scales*, Taylor and Francis, London.

Pheng, K.S., 1996. Putting GIS into practice in developing countries. In: *Proceedings of the Conference on GIS and Developing Countries: The Practice of Applications* held at Utrecht, the Netherlands, August 2-3, 1996.

Tewari, V.K., 1996. GIS applications in rural development planning and management in developing countries. In: *Proceedings of the Conference on GIS and Developing Countries: The Practice of Applications* held at Utrecht, the Netherlands, August 2-3, 1996.

UN Economic and Social Commission for Asia and the Pacific - ESCAP, 1994. *Space Technology and Applications for Sustainable Development in Asia and the Pacific: A Compendium*, United Nations, New York.

UNEP and WCMC, 1996. *A Guide to Information Management* (in the context of the Convention on Biological Diversity).

Valenzuela, C.R., 1996. Geographic Information Systems in natural resources management. In: *Proceedings of the Conference on GIS and Developing Countries: The Practice of Applications* held at Utrecht, the Netherlands, August 2-3, 1996.

Westley, F., 1995. Governing design: the management of social systems and ecosystem management. In: Gunderson, L., C.S. Holling & S. Light, (eds.) *Barriers and Bridges to the Renewal of Ecosystems and Institutions*, 391-427.

Index

T - #0566 - 101024 - C0 - 218/152/10 - PB - 9780815382584 - Gloss Lamination